The Simple Modem Book

by Bill Potts

alpha books

A Division of Prentice Hall Computer Publishing
11711 North College Avenue, Carmel, Indiana 46032, USA

Dedication

To Gerry Hinchliffe, an outstanding teacher who encouraged me in my somewhat difficult formative years.

 For information, address Alpha Books, 11711 N. College Avenue, Carmel, Indiana 46032.

International Standard Book Number: 1-56761-146-X
Library of Congress Catalog Card Number: 93-83121

95 94 93 7 6 5 4 3 2 1

Interpretation of the printing code: the rightmost number of the first series of numbers is the year of the book's printing; the rightmost number of the second series of numbers is the number of the book's printing. For example, a printing code of 93-1 shows that the first printing of the book occurred in 1993.

Screen reproductions in this book were created by means of the Collage Plus program from Inner Media, Inc., Hollis, NH.

Printed in the United States of America

Publisher

Marie Butler-Knight

Associate Publisher

Lisa A. Bucki

Managing Editor

Elizabeth Keaffaber

Acquisitions Manager

Stephen R. Poland

Manuscript Editor

Barry Childs-Helton

Cover Designers

Jay Corpus, Kathy Hanley

Designers

Amy Peppler-Adams, Roger Morgan

Indexer

Jeanne Clark

Production Team

Diana Bigham, Katy Bodenmiller, Julie Brown, Scott Cook, Tim Cox, Terri Edwards, Mark Enochs, Linda Koopman, Tom Loveman, Beth Rago, Joe Ramon, Carrie Roth, Linda Seifert, Greg Simsic, Mary Beth Wakefield, Kelli Widdifield

Special thanks to C. Herbert Feltner for ensuring the technical accuracy of this book.

Some illustrations in this book created with clip art from CorelDRAW! 3.0 from Corel Corporation.

Handy Tables, Checklists, and Exercises in This Book

Table of Contents

Introduction

We are no longer on the threshold of the Information Age. We have stepped inside and locked the door behind us. There is no going back.

Living in the Global Village

In 1968, the late Canadian philosopher and guru Marshall McLuhan gave us the term *Global Village*. McLuhan's characterization of the (developed) world, based on the availability of television and telephones, may have been a little premature. But thanks to personal computers, modems, and online services such as CompuServe and PRODIGY, McLuhan's Global Village—in which we would unite and interact through electronics—is now a reality.

Today, there are computers in tens of millions of homes throughout the developed world. Some are in home offices; some are used for game playing; some are used for personal correspondence, checkbook balancing, home recipe storage and retrieval, and so on. In ever-increasing numbers, they are also used to communicate with services outside the home.

Making Modems Simpler

Giving your computer the ability to communicate requires a *modem*. (The term is a contraction of modulator/demodulator, one of the things we explain in this book.) You may already be familiar with modems, or you may view them with some apprehension. If you are in the latter category, don't despair. The manufacturers of personal computers, software, and modems compete for your business by making things as simple as they can (usually successfully—but not always). With their products and this book, you and your computer will be communicating in no time.

Is This Book for You?

If you decide you want to be a PC user with the right connections, Chapter 1 provides some help with this decision. If you don't feel as comfortable with technical matters as you would like, then you will certainly find this book helpful.

Once you have made your decision to get started with modems, "going online" is just a matter of following a few straightforward procedures. When you select your modem, you may make some nonrational (that is, emotional) decisions, like choosing the best-looking modem from a range of otherwise similar (and similarly-priced) products. Installing and operating your modem—and, if necessary, performing basic troubleshooting—will require only that you follow simple instructions, which are conveniently compiled in this book.

But I'm No Good with Tools!

(If this doesn't apply to you, see the next heading.)

You won't even need tools to install some modems. For others, you may require a very small, conventional screwdriver to secure the cable connectors firmly to the modem and your PC. If you decide on an internal modem, you will have to open up your personal computer to install it (as you would to install any other add-in board). In that case, you may need a conventional screwdriver, a Philips screwdriver, or a small nutdriver (for screws with hexagonal heads). You can open some PCs with a dime (IBM PS/1s and PS/2s, for example); some have screws with long, knurled heads, which can be loosened and tightened with your finger and thumb. In short, you don't need to be familiar with any complicated tools to get the job done.

But I AM Good with Tools!

Okay, okay! You just won't be too challenged by this project (installing your modem)—which will leave you with more time for all your other projects.

Let's Get On With It

Before you jump in, take a minute to see how the book is organized. Don't worry if you're not familiar with some of the terms I'm about to throw at you. The chapters explain the terms you need to know. (If you have to know now, you can check out the Glossary, too.)

PART I, GETTING THERE, takes you from the decision to buy a modem to completing its installation. This part consists of the following nine chapters.

Chapter 1, "Why Do You Need a Modem?," helps you answer the question posed in the title. In particular, it introduces you to the staggering range of information and services available to users of modem-equipped personal computers.

Chapter 2, "What Exactly Is a Modem?," will help you if you are not completely sure of what a modem is. Even if you already know the obvious—that it's a communication device—you should find this chapter informative and interesting.

Chapter 3, "Buying the Right Modem for You," deals with the sticky question of choosing a modem from the bewildering array of types, makes, and models. Although it doesn't recommend one manufacturer over another (this isn't *Consumer Reports*), it will help you with the choice.

Chapter 4, "What About the Software?," reveals what you suspected all along: in addition to the modem, you're going to need some software. Here you will learn what is included when you subscribe to the better-known online services, the minimum you need for Captain Billy Whizbang's local bulletin board, and the kind of software you can expect to be packaged with the modem itself.

Chapter 5, "The Fax of Life," revisits the hardware to explore what has become a very common modem option—fax capability. We look at the pros and cons of using your computer as a fax machine.

Chapter 6, "Planning Ahead," addresses an often-neglected step—planning your modem installation before you actually go out and spend your hard-earned money. You may want to make some changes to your phone installation. Don't worry, though. If you don't mind tying up your existing phone line whenever you use your modem, you don't absolutely have to make any changes.

Chapter 7, "Where to Buy a Modem," finally helps you get around to the actual purchase. In particular, it addresses the questions of where and how to buy both the modem and the software.

Chapter 8, "Installing Your Modem," takes you through the step-by-step physical installation procedure, including an admonition that, however helpful this book may be, you should always remember to *read the manual*.

Chapter 9, "Getting It Running," deals with the moment of truth, the peak experience—installing the software and establishing that first live connection. Eureka!

PART II, GETTING AROUND, recognizes that you are now a suave, sophisticated, and reasonably laid-back consumer of communication services—or at least would like to become one. Now that you're hooked (or at least hooked up), you can start to explore some of the capabilities that moved you to buy the modem in the first place. Of course, you don't really need to wait until you buy your modem to read Part II. It contains a great deal of information that will be helpful in making your original purchase decision.

Chapter 10, "Exploring PRODIGY," introduces you to this well known and sometimes controversial online service, owned jointly by IBM and Sears-Roebuck. PRODIGY is probably the easiest online service to use. Although it also bombards you with ads for merchandise and services, advertising allows PRODIGY to provide you with unlimited connect time with no hourly charge (except for what PRODIGY calls "plus" features)!

Chapter 11, "CompuServe," describes the service whose subscribers are sometimes inclined to consider themselves in a somewhat higher "techie" class than the PRODIGY subscribers, although many of them are also closet PRODIGY users. CompuServe has responded to PRODIGY's free connect time by way of a package deal, for which connect time charges are waived for a number of basic services (news, weather, sports, travel, shopping, and so on).

Chapter 12, "America Online," gives you a glimpse of the new kid on the block. Until 1991, this service was available only to Apple Macintosh users. Now, users of IBM-compatible personal computers have a choice of two software packages for access to America Online—one that looks a bit like Microsoft Windows, and one that *is* a Windows 3.1 application.

Chapter 13, "Other Electronic Mail Options," calls your attention to the online services whose major function is E-mail. (Electronic mail allows you to send messages to and receive messages from those who subscribe to the same online service as you, and even those who subscribe to other online services.) Like MCI's Friends and Family option for its long distance customers, the value of services like AT&TMail, MCI Mail, and SprintMail depends on whom you want to talk to. You'll also learn about some of the extra mail options available on CompuServe, MCI Mail, and other services.

Chapter 14, "Bulletin Boards," returns to a subject touched on in Chapter 1. A bulletin board service allows you to "post" messages under one or more topic headings, to read messages posted by others, and to reply if you wish. The real modem junkies sign on to some of these well into the night.

Chapter 15, "Transferring Files," discusses the general topic of electronic file transfer, either between two personal computers or when using one of the major online services. The standard or optional software offered by online services makes file transfer easy. For other situations, such as direct file transfers, you'll need to know about some of the available choices.

Chapter 16, "Electronic Bill Paying," talks about the bill paying options available to you through the major online services or directly from your bank.

Chapter 17, "Remote Computing," may be of interest if you want to telecommute—work on your computer at the office while sitting comfortably at your computer at home.

Chapter 18, "Modem Care and Troubleshooting," provides some common-sense advice for protecting your modem from dust and damage, plus some tips on troubleshooting.

The Appendixes provide some additional information better expressed as tables or lists.

Appendix A, "Modem Indicator Lights and What They Mean," guides you through that bewildering array of blinking lights you'll see on an operating modem.

Appendix B, "The Hayes AT Command Set," lists some of the important commands that may be sent from your PC to a Hayes or Hayes-compatible modem. Your modem manual will tell you which ones are supported on your modem.

Appendix C, "Glossary," defines terms used in this book, plus a number of other data communication terms you might encounter. As befits a glossary, it's a little more formal in tone than the rest of this book.

Appendix D, "Modem Shopping Checklists," helps you ensure that you don't overlook anything.

Conventions Used in This Book

Whenever we want to reinforce the message of this book, we do so with one of several icons.

> **WORD POWER**
>
> **WORD POWER** draws your attention to a term you must understand before you continue. A definition may be repeated (or reworded) in later chapters. Most terms defined this way also appear in the Glossary (Appendix C), but with more formal wording.

COMMON SENSE

> **COMMON SENSE TIP** highlights practical advice, such as never paying list price.

BEST BET TIP provides you with a general recommendation for those situations where you are faced with a choice.

PITFALL TIP indicates something to avoid, such as paying for a useless feature. It also indicates something to which you should pay particular attention—for example, remembering to unplug your computer before opening it up.

TRY IT! exercises show you how to practice a procedure I'm discussing, or let you test how well you understand what you've learned.

Acknowledgements

I would like to acknowledge and thank the following people and organizations for their assistance and cooperation.

America Online, Children's Television Workshop, and Prodigy Services Company, for permission to use screen shots.

Digicom Services, Inc., for providing information in such a cordial and helpful manner.

The editorial staff of Alpha Books, for helping to turn so many of may original manuscript's byways into highways, and for giving me some breathing room in the schedule.

I would also like to acknowledge the passive, but vital role of *PC Magazine*, which is such a good source of current information.

Bill Potts, April 1993

CHAPTER 1

Why Do You Need a Modem?

Why, indeed? It's certainly something you can live without, if you have to. After all, some people live without a TV set, some without a radio, some without even a telephone.

However, your decision to buy (or borrow) this book indicates that you probably haven't decided to take refuge in a Tibetan monastery, you probably have (or plan to buy) a personal computer, and you have at least heard of modems.

> **WORD POWER**
>
> A *modem* is a device that is connected to both your computer and your telephone line, allowing your computer to transmit data to (or receive data from) a similarly-equipped remote computer.

Learning more about modems may not make you realize you need one, but the variety of things you can accomplish with a modem may make you want one—and soon. Let's look at a couple of basic facts now. The details about what a modem is can wait until Chapter 2.

What's a Modem For?

Until fairly recently (about 1991), you could say that a computer's relationship with a modem was very similar to your own relationship with your telephone. It provided a means of two-way communication, but for text (typically) rather than speech.

The emphasis on text reflected the technology of the time. The typical modem for a personal computer could transmit the equivalent of 120 characters of text per second—which beats a fairly good typist by a factor of about 20 (and without coffee breaks)—but still is not blindingly fast. Today, however, we are not content with transmitting words; we want to send and receive pictures, too. After all, we receive an enormous amount of information in the form of images—on TV, in the print media, by fax, and now (if you and at least one relative want it, and can afford it) by video phone.

Personal computers have kept pace. They have always presented information visually, first in text form, and then by means of color graphics of ever-increasing resolution. Recently they have come full circle with the addition of sound—stereophonic, of course, with voice, music, and special effects. These capabilities have been exploited to the greatest extent in computer games, many of which now have full-motion video and sound effects in all their cacophonous splendor.

Graphics and digitized sound require an enormous amount of internal memory (often referred to as *RAM*, or *random-access memory*) and an enormous amount of space on your hard disk for file and program storage. In addition, they have formed the foundation of a market in which the files (programs, graphic images, digitized audio, and so on) are delivered electronically—that is, via modem (preferably a fast one).

Specific Things You Can Do with a Modem

Now you know that modems can transmit graphic and text files electronically. But what does that mean to you in the real world? Actually, it may mean a lot.

With a modem, you can use your computer for an enormous number of additional purposes—some practical, others merely amusing or entertaining. Many have wide appeal; some are more specialized. This section summarizes just a few, and gives you some idea of the variety available. Part II of the book deals with several at greater length.

Exploring an Online Service

You can use your modem to dial into an *online service*, which can best be described as a publicly-accessible information utility. Like the phone company (and unlike gas, water, and electricity utilities), it operates in both directions, allowing you to not only receive information, but send it too.

Each online service is, in fact a cornucopia of services, available for a monthly fee (plus extra charges, in some cases) to every personal computer user who has a modem.

Figure 1.1 shows the initial *Browse* menu for CompuServe, one of the leading online services. This menu is displayed by the CompuServe Information Manager (a software package I discuss in detail in Chapters 4 and 11). The departments listed in this menu are only the tip of the iceberg. Selecting any of them will lead you to a vast array of choices, allowing you to engage in such activities as:

- Selecting and ordering merchandise.
- Sending and receiving electronic mail (discussed later in this chapter).
- Exchanging ideas with others in a large number of fields of interest.

- Searching reference libraries and online encyclopedias for information to satisfy your curiosity, or to complete a business or educational project.
- Checking current prices and trends for stocks and other securities, and even handling your own stock and bond trades.
- Reporting problems on (and getting fixes for) many of the software packages you run on your personal computer.
- Getting advice from both experts and non-experts on a variety of issues, including computer software.
- Keeping up to date on the latest in news, weather, sports, politics, art, entertainment, and so on.

As with the menu in Figure 1.1, this list has only scratched the surface. Chapters 10, 11, and 12 will provide you with more specific information (including pricing) about the three leading online services—PRODIGY, CompuServe, and America Online.

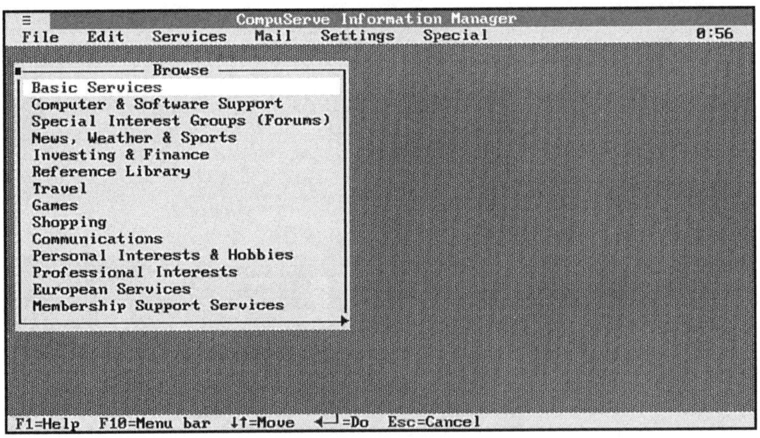

Figure 1.1
The Browse menu for CompuServe.

Sharing Information on a Bulletin Board System (BBS)

Bulletin board systems are small online services, usually operated by one person, a couple of people, or a special-interest group. Quite obviously, they are unable to provide the kinds of services offered by the major online-service companies, but they do provide forums for the exchange of ideas, often about local issues.

Some BBSes operate as high-tech lonely-hearts clubs, some are hotbeds of social activism, some distribute computer games (for a fee), some handle classified advertising, and some even distribute quite racy materials (after checking that you are a bona fide adult).

Chapter 14 tells you more about BBSes and how to find out about those in your area.

Sending and Receiving Electronic Mail

> **WORD POWER**
> *Electronic mail* is a message (or messages) sent electronically from one user to one or more other users on an online service, bulletin board system, or computer network. Usually abbreviated to *E-mail*.

For many people, E-mail is replacing the Postal Service (and at much lower cost). Suppose you're already using your computer for correspondence. You probably load your trusty word processing software, compose your letter, save it on disk, print it out, address an envelope (using your computer, typewriter, or pen), seal it, put a stamp on it, take it to a mailbox or post office, or give it to your mail carrier.

Even if you don't put a value on your labor, you have certainly incurred material costs, including the stamp, of 29 to 35 cents (or more, if the rates have increased since this was written). Depending on the destination, delivery may take from a day to a week. If the letter is going overseas, the stamp cost is almost double, or even more if you are a prolific writer and the weight exceeds half an ounce.

Consider one of the choices you have with a modem. You load the CompuServe Information Manager (CIM), one of the applications described in Chapter 11. You select the **Mail** menu and choose the **Create Mail** option. You enter a subject heading and the addressee's CompuServe address (which you can retrieve from an electronic Address Book maintained by CIM with your help). Then you compose your message as you would an ordinary letter, save it in an electronic File Folder (so you have a record of it), then select the **Send** option. Using your modem, CIM dials the local CompuServe access number and sends the message, which the addressee receives the next time he or she connects to CompuServe. To retrieve incoming mail, all you do is choose the **Get New Mail** option of the **Mail** menu. Again CIM does the dialing and manages the connection for you. Your CompuServe subscription includes a generous monthly E-mail allowance at no extra charge.

PITFALL! In general, the person you want to send E-mail to must also subscribe to the online service you're sending the mail to. Your online service may also enable you to send E-mail messages to other users. If you want to use your online service primarily for E-mail, choose the service that lets you send messages to the greatest proportion of your recipients.

CompuServe, MCI Mail (see Chapter 13), and other online services even let you send E-mail to a postal address (in case the person you're writing to is not a subscriber). Mail to someone in England, for example, will be printed out at a location in England, placed in an envelope, and mailed locally. There is, of course, a charge for this kind of service.

Many online services, including CompuServe, PRODIGY ®, and America Online, provide a means to easily send E-mail to political candidates, state and national legislators, activist organizations, and special-interest groups.

Most of them also allow you to send mail to subscribers on other online services and E-mail systems.

Talking Directly to Other Computer Users

Local BBSes often provide the means for having live conversations with other users. If you sign onto any of CompuServe's special-interest forums, you can initiate a conversation with others who are currently signed on. Most users are inclined to wait for *conferences* to be scheduled on a specific topic, and at an announced time; Figure 1.2 shows a conference in progress.

You can also connect directly with other users (skipping the online-service connection), and use the *chat feature* of your communications software to hold one-on-one sessions. All you need to do is have your modem dial the phone number that the other modem is connected to.

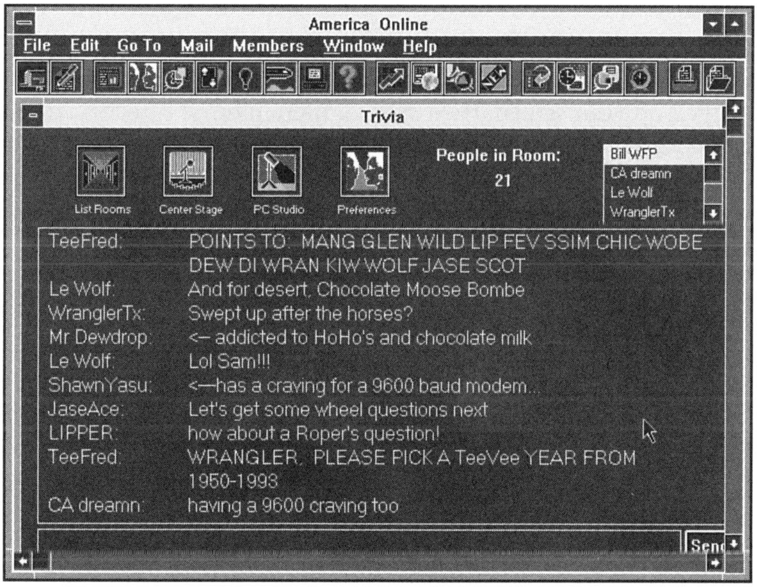

Figure 1.2

Conversation in one of America Online's many "conference rooms."

Faxing with a Modem

Many modems sold today are described as *data/fax modems*. Does this mean you can actually use your computer to send and receive faxes? It certainly does. Such modems incorporate the same technology used in the send-and-receive circuits of fax machines. With the right software, you can use your computer to generate documents for faxing, and to display or print incoming faxes.

Using your computer for faxing may or may not be a good idea for you. After you've read Chapters 5 and 17, you will be better prepared to make your own decision about this.

Electronic Malls

No, this isn't a misprint. You can shop for anything from computers to sexy lingerie right at your computer.

PRODIGY, a low-cost online service targeted for novice/home users, makes shopping especially easy. You can select a supplier from a directory displayed on your screen, then select a merchandise category, then an item, then place an order. In many cases, you will get to see a picture of the item (or, rather, an artist's rendition) before you make your decision, as shown in Figure 1.3. PRODIGY's simple procedures have safeguards to make sure you don't place an order inadvertently.

COMMON SENSE

If you have a separate phone line for your modem, you can call one or two local stores while browsing on PRODIGY, just to compare pricing.

PITFALL!

If your sales resistance is zero, stay away from PRODIGY; this service derives much of its revenue from advertising, which appears at the bottom of your screen every time the screen is updated. You can view further information on the advertised product immediately, place an order if you wish, then return to what you were doing when the advertisement appeared.

Figure 1.3
Some electronic malls, such as PRODIGY's, show you what an item looks like before you place your order.

Tracking Your Investments

You can use any of the major online services to track the stocks in your portfolio. You can opt for the standard quotation service, which gives you prices that are almost current—that is, they are subject to a minimum delay of fifteen minutes. If you are a "serious investor," you can opt for an extra-cost service, in which the quoted prices are up to the minute. Still further extra-cost options allow you to trade stocks online.

Both PRODIGY and America Online allow you to save a list, by stock market symbol, of all the stocks in your portfolio (or, at least, a reasonable number of them), allowing you to obtain quotes on all of them at the same time. America Online's StockLink option even allows you to include, for each stock, the number of shares you own and, if you like, what you paid for them. When you request quotes, you get a full and current valuation for your portfolio. You can even generate stock performance charts.

PRODIGY's extra-cost Strategic Investor option allows you to research and evaluate over 5000 securities.

COMMON SENSE

> If you really are a serious investor, compare the monthly and transaction fees for the online stock trading services with the commissions you pay your broker. If you don't rely on your broker for advice on buying and selling, you might want to take advantage of the lower cost of do-it-yourself trading.

Contests

If you're a contest junkie, you'll be happy to know that you can get your fix all day long on America Online, CompuServe, or PRODIGY.

Fun for the Kids

Do you have any kids (or grandchildren) who are six years old or less? Do you watch Sesame Street with them? For an extra monthly fee, PRODIGY will provide you and your toddler with interactive games, quizzes, stories, and so on. All the Sesame Street characters are there, beautifully drawn and in full color (you wouldn't want to use a monochrome monitor for this one). Figure 1.4 will give you some idea of the available fun.

Politics and Other Social Issues

PRODIGY's Close-Up forum is a sounding board for many of the controversial issues of the day. Is the President doing a good job? Is there a justification for increased taxes? Should we have a school voucher system? Should abortion be banned? Restricted? Unrestricted? You will find hundreds of postings by PRODIGY subscribers on these and other topics, from the left, center, and right of the political and social spectrum. You can reply to other people's comments, or post your own opinions. You can print or save to disk anything you consider memorable.

Figure 1.4
A "Name That Muppet" contest on PRODIGY.

What's Free and What's Not

You usually get what you pay for. Or, to be more precise, you're unlikely to get what you don't pay for.

Obviously, you're going to have to pay for your modem (unless you can convince a friend or relative to buy it for you). You'll probably also have to pay for the software needed to take full advantage of it. But what about monthly and other fees for the online services you might want to use?

The large online services, like CompuServe, PRODIGY, and America Online, charge a basic monthly rate. PRODIGY includes unlimited use in the monthly rate, although they do charge for electronic mail after the first thirty messages per month, and have recently added some extra-cost services. The penalty you pay for the unlimited use comes in the form of advertising—every time your screen is updated.

CompuServe offers a very low base rate ($2.50 per month) plus a connect-time fee. There is also a higher (but still reasonable) base rate of $8.95 per month, with no connect-time fee for a specified set of basic services, including a generous allowance for electronic mail. Extended services, which they identify with a plus sign (+), are subject to a connect-time fee. Also, a number of organizations offering services via CompuServe charge for such services, usually by the transaction. A dollar sign is used to warn you whenever you select a service for which a charge applies.

Both PRODIGY and CompuServe support a range of modem speeds. PRODIGY, whose normal monthly fee is $14.95, has a small monthly surcharge of $2.00 for the fastest modems (9600 bits per second). CompuServe charges twice as much for connect time at 9600 bits per second ($16.00 an hour) as it does at 2400 bits per second ($8.00 an hour). Depending on how you use CompuServe, this can be a real bargain.

WORD POWER

Bits per second is a measure of the rate at which information is transmitted over a communication link. The standard abbreviation is *bit/s*. A common abbreviation is *bps*. See also Chapter 2 and the Glossary (Appendix C).

America Online's base monthly rate of $9.95 includes five hours of free connect time; as of July 1, 1993, charges of $3.50 per hour start once the free time is used up. So far, they don't offer service at speeds higher than 2400 bits per second; however, as of this book's publication date, they have announced plans for 9600-bit/s service.

MCI Mail charges a base monthly rate of $10.00, nothing for connect time, and a fee for each message. It also offers some additional, chargeable services.

Chapters 10 to 12 include pricing information for PRODIGY, CompuServe, and America Online. It is, of course, subject to change.

The charges for the local bulletin boards (see Chapter 14) vary from nothing to very high. The ones run by amateurs will often accept voluntary contributions to help defray the cost of operation. The others may charge a monthly fee, by connect time, or a combination. Those offering digitized art, games, and other software will bill you for whatever you download to your PC.

Ready for the Next Step?

Now that you know something of the treasures in store for PC users with modems, you're ready for Chapter 2, which should help you understand what a modem is, what it does, and a little about how it does it. In the meantime, you should now know the following:

- A modem is a device that allows computers, terminals, and so on to communicate with one another.
- Online services and bulletin board systems (BBS) let you perform a number of tasks electronically, including sending and receiving mail, shopping, researching any topic you want, playing games, and more.
- You can also use your modem to chat or exchange files with another modem user, or (if you have a fax modem) to send and receive faxes.
- The three best-known online services are CompuServe, PRODIGY, and America Online.
- E-mail is a faster (and often lower cost) alternative to the Postal Service.
- The speed of a modem is expressed in bits per second.

CHAPTER 2

What Exactly Is a Modem?

A *modem* is a device that is connected to both your computer and your telephone line, allowing your computer to transmit data to (or receive data from) a similarly equipped remote computer. (See Figure 2.1.)

Internal and External Modems

Modems for personal computers are available in two basic forms: internal and external. You can connect an *external* modem to one of your computer's *COM ports* (outlets on the back of your computer) by means of a cable. It also needs to be plugged into an electrical outlet.

Internal modems connect to the telephone line in the same way as external ones. The difference is that they are mounted on a card and installed in one of your computer's *expansion slots* (places for circuit boards within your computer). The modem on the card becomes part of your computer.

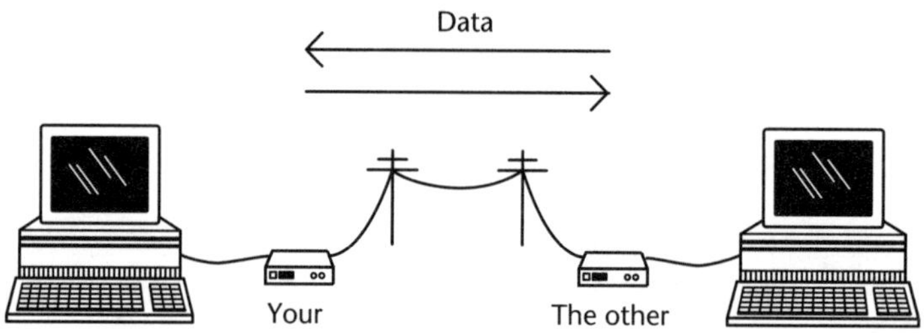

Figure 2.1
Modems transmit data from your computer to another computer over phone lines.

Figure 2.2 illustrates an internal and external modem. Both work the same with your communications software. For more about the difference between internal and external modems (and which you should consider buying), see Chapter 3.

Internal Modems

The typical internal modem comes in the form of a circuit board made to fit into one of your computer's expansion slots. Its backplate has two sockets. One of the sockets is for the cable to the telephone wall socket; the other is for the cable from your telephone (if your phone and modem are sharing the same line, and if there is only one wall socket). The sockets and the plugs that go into them are known as *RJ-11* sockets and plugs (or sometimes by the longer names "RJ-11 modular socket" and "RJ-11 modular plug").

The backplate may, additionally, have some *LEDs* (light-emitting diodes) to indicate *status* (power on, transferring data, error, and so on). It may also have one or more switches, or an access hole for board-mounted DIP switches. Figure 2.3 shows the backplate of a typical internal modem.

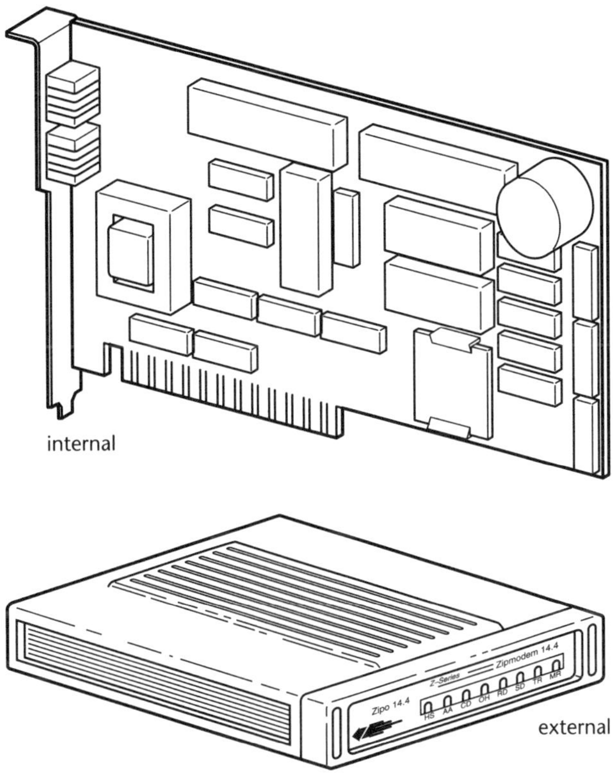

Figure 2.2
Two forms of ***modem****: internal and external.*

External Modems

Today's technology ensures that the biggest external modems (especially newer ones) are only about the size of this book. The smallest ones are not much bigger than a pocket calculator.

External modems usually have four connectors—two RJ-11 modular sockets, a 25-pin female connector, and a power connector. The RJ-11 sockets serve the same purpose as on the internal modem. The power connector supplies power to the modem via a cable from a transformer (included in the modem package) that plugs into a wall socket. Some manufacturers put the transformer inside the modem, which lets it use a conventional AC power cable and plug. Some modems, such as the Hayes Personal Modem 2400, plug directly into the wall socket. Figure 2.4 shows the connectors.

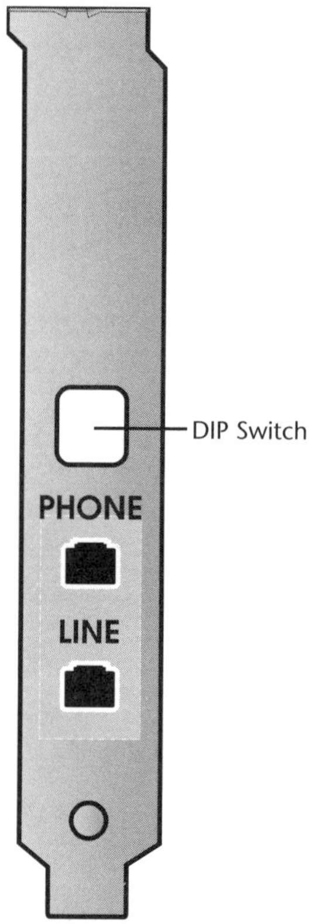

Figure 2.3
Internal modem backplate.

Figure 2.4
An external modem with connectors.

Modem manufacturers almost always supply a cable, so you can connect the modem to one of your computer's COM ports. The cable has a connector on either end—one for the modem (25-pin male), and one for the COM port (9-pin female). If you have an early IBM PC (pre-"AT" model), and notice it has one of the old 25-pin COM ports, don't despair. Often a special short cable comes with the modem, and allows you to connect the modem's 9-socket cable to the old 25-pin port. Figure 2.5 shows this connector.

WORD POWER

A *port* is a point at which another device (modem, printer, etc.) can be connected to your computer. Typically, your computer has two *serial ports* (COM1 and COM2) and one *parallel port* (LPT1—normally used for a printer). Connection to each serial port is made via a 9-pin (or sometimes 25-pin) male connector at the back of the computer. Connection to a parallel port is made via a 25-pin female connector, also at the back of the computer. A serial port can handle data transfer (to or from your computer) bit by bit; only two pins of its connector are used—one to send, and one to receive. A parallel port can send all eight bits of a byte to the printer at the same time; as you might expect, eight of the 25 pins are used for this.

Note that it is normal to refer to both male and female connectors as *9-pin* or *25-pin*, even though the female connector has sockets rather than pins.

PITFALL!

Some modem manufacturers save on both cost and space by building modems with permanently attached modem cables. This eliminates two connectors—the one on the modem, and the one at the modem end of the cable. Other manufacturers reduce the space requirements by using a compact (but nonstandard) connector on the modem, and on the cable at the modem end. If your dog chews through either of these cable types, the replacement may be obtainable only from the original manufacturer. For the permanently attached cable, you may have to pay for the installation of the new one (unless you're really handy with a soldering iron).

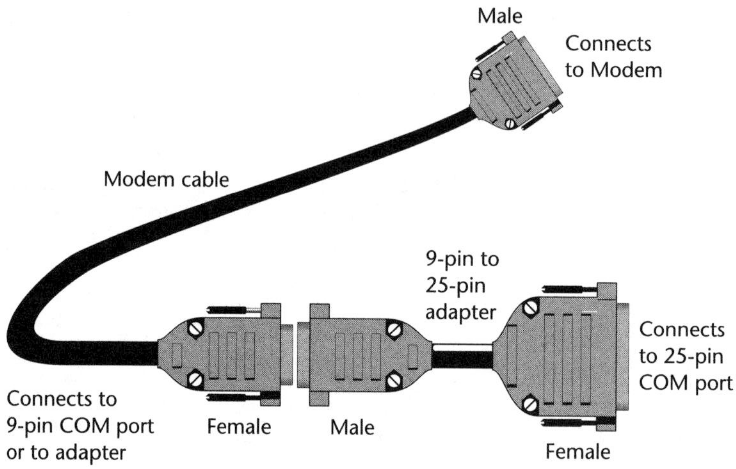

Figure 2.5
A modem cable and a 9-to-25-pin adapter cable.

BEST BET

If you're only going to use your modem with one personal computer, you expect to keep both the computer and modem for another year or two, and you have an available slot, then consider an internal modem. If you want flexibility, get an external modem—but remember, you also have to plug it into a power socket.

The Interface

If you have an external modem, it will be connected to one of your computer's COM ports by a cable. The connection made by the cable between the COM port and the modem is called the *DTE-DCE interface*. The computer is the DTE (data terminal equipment), and the modem is the DCE (data circuit-terminating equipment).

To make it possible for you to buy a modem off the shelf (and be assured it will work with your computer), there must be standards defining the

DTE-DCE interface. Such standards define the circuits connected by the cable, and what kinds of electrical signals they are to carry (see "Zeros, Ones, and COM Ports," later in this chapter). The best-known standard—and the one which applies to your computer, as well as to the kinds of modems described in this book—was defined some years ago by the EIA (Electronic Industries Association). You'll see it referred to as *RS-232* or *RS-232C* in advertisements, in modem documentation, and on modem packages. (Actually, this terminology is more than six years out of date. The standard is now correctly written as *EIA-232-D*. Some manufacturers, including IBM, use the up-to-date term.)

If you're interested in what the circuits are, see the Glossary entries for CTS, DCD, DSR, DTR, RD, RI, RTS, and SD. Appendix A describes the modem indicator lights that might be associated with each of the circuits. Chapter 19 tells you, step by step, how the lights (therefore the circuits) might behave when the modem is in use.

Does This Ring a Bell?

You may be wondering why your computer can't talk directly through the phone lines. That's a good question, of course. Before I give you the answer, let's consider what your telephone is designed to do. Today's telephones are just a refinement (although a considerable one) of the instrument Alexander Graham Bell first used to summon his colleague, Mr. Watson.

The microphone in the mouthpiece converts sound waves to electrical signals, and the speaker in the earpiece converts electrical signals to sound waves. The electrical signals are in the form of voltage fluctuations. These represent the frequency of the sound waves, and the loudness (amplitude) of your voice at any instant. We refer to such signals as an *analog signals*, because they are an electrical analogy of the sound of your voice. Like sound waves, the electrical signals fluctuate constantly (see Figure 2.6). The link between your telephone and the phone company's switching center can carry only this kind of signal.

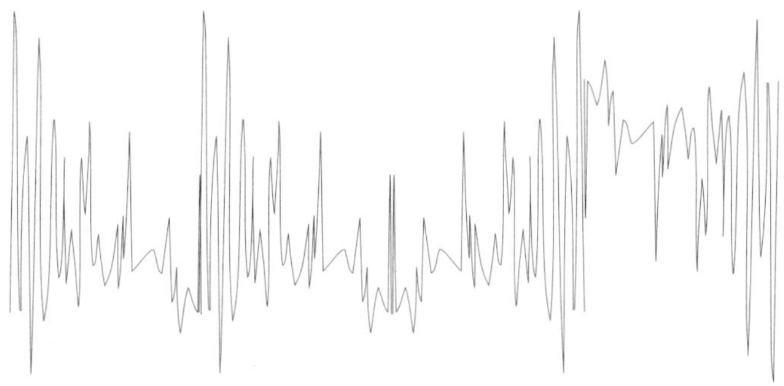

Figure 2.6
Your telephone sends signals like this one over the phone lines.

Zeros, Ones, and COM Ports

Now let's consider how your computer talks to the outside world. You probably know already that your computer's most basic unit of data is the *bit* (binary digit), which can represent either of two values—0 or 1. You probably also know that your computer works with groups of eight bits, and that such a group is called a *byte.*

Your computer communicates with other devices by transmitting (or receiving) a series of bytes. It can do this byte by byte—over a parallel printer connection, for example—or bit by bit, through a serial communication port (COM port).

Like the telephone, a COM port uses electrical signals. Unlike the telephone, it uses positive and negative voltages (typically +12 V and –12 V) to stand for either zero or one. The voltage is positive if the bit has a value of zero, and negative if it has a value of one; each bit will be one or the other. Figure 2.7 shows the voltage changes for the transmission of the bit string 1001011010.

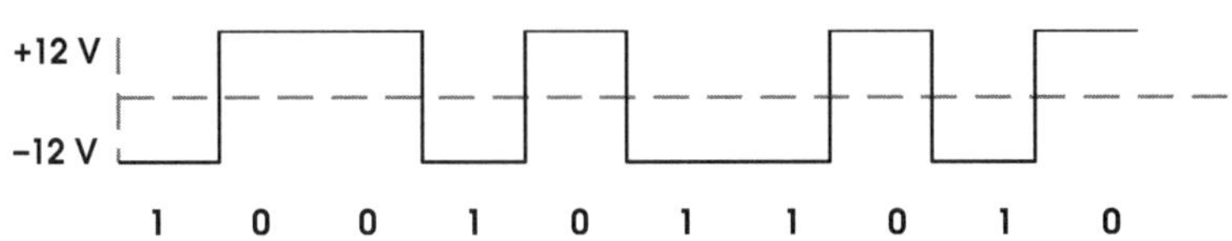

Figure 2.7
How bits are represented electrically. Your computer represents each bit with a positive or negative voltage. This digital signal is very different from the analog signal your phone uses.

We refer to the signal at the COM port as a *digital* signal. Digital signals are extremely easy for your computer to work with, because every signal can be interpreted as a positive or negative. (I won't bother you with all the details of the COM port here. Those are covered in Chapter 8 and in the Glossary.)

What a Modem Does

One of the older (and simpler) types of modem converts the negative and positive voltages to audible tones. A negative voltage (representing a 1 bit) is converted to a tone of one pitch; a positive voltage (representing a zero bit) is converted to a tone of a slightly lower pitch. The receiving modem converts the tones back to the original negative and positive voltages.

In Figure 2.8, I've illustrated the tones in two ways, along with the digital signal you saw in Figure 2.7. If you like graphs, you'll observe that the wavy line is oscillating more rapidly (that is, at a higher *frequency*, or pitch) for the ones than it is for the zeros. If you can read music, the two-note "Tone Poem in D Major" may appeal to you a little more. At 18,000 half-notes per minute (300 per second), its performance on a musical instrument may present some problems.

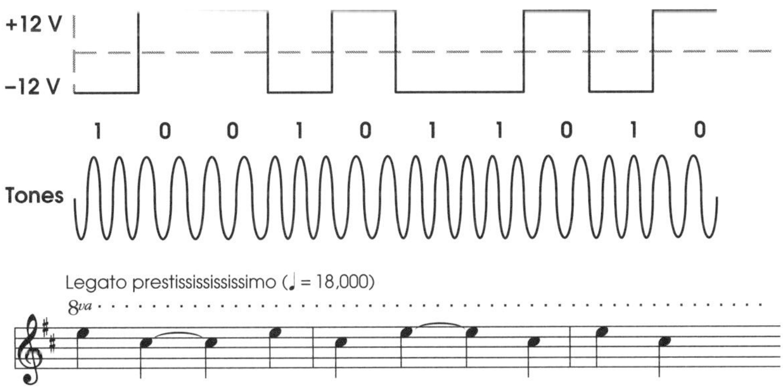

Figure 2.8
Two ways of looking at very simple modulation.

This use of tones of two frequencies is called *frequency-shift keying*. It was used by the old Bell 103 modems, and in Europe by CCITT V.21 modems (see Glossary). Both types were limited to speeds of no more than 300 bits per second.

Frequency-shift keying is very inefficient, especially where only one tone is used at a time. Today's modems use more than two possible *states* (each of the two tones in our example represents a "state"). These more advanced modems may combine states (such as amplitude) with changes of state (such as phase change) to represent groups of two, three, four, or even more bits. (I'm not going to try to explain what "phase change" is. It's sufficient to say that it's something a modem can do to the signal it's sending, and which the receiving modem can detect.)

Whether it's simple or complex, we call the process by which a modem converts digital signals into analog form *modulation*, and we call the reverse process *demodulation*—hence the term *modem* (modulator/demodulator).

We don't need to understand the details of the various kinds of modulation, but the grouping of bits just mentioned leads to our next topic—the baud.

Baud and Bits Per Second—Not the Same

Despite its sound, *baud* is not a subject of limericks or bawdy barroom ballads. It is, rather, a unit of measure—named after one of the pioneers of the printing telegraph (Emile Baudot), and equal (very simply) to one *change of state* per second.

Today's telephone connections can be used reliably to transmit a signal that changes state up to 2400 times per second. Such a connection is operating at a rate of 2400 baud.

With complex modulation techniques, we can have not just two states, but many states. With eight states, we can use each state to represent one of the eight possible groups of three bits (000, 001, 010, 011, 100, 101, 110, or 111). So, at a signaling rate of 1600 baud (for example), we can transmit 4800 bits per second. Today's fastest modems use 64 states, which can represent all the possible values of a group of six bits. At 2400 baud, this works out to 14,400 bits per second.

Clearly, *baud* is not the same thing as *bits per second* (which we'll discuss next).

PITFALL!

> Many modem advertisements give the modem's speed (incorrectly) in baud. They really mean *bits per second* (bps). A 9600-baud modem is impossible, because good-quality phone lines can only handle up to 2400 baud. As you have just learned, today's modems (with their complex modulation techniques) can transmit and receive 9600 bits of information per second (and more) while signaling at only 2400 baud.

What About BPS? Or Is It bps? Or What?

Actually, it's "or what!" Almost all the standard organizations use *bit/s* as their accepted abbreviation for "bits per second." *BPS* and *bps* are both nonstandard ways of writing the same thing, but they are very commonly used, so you will find "bps" in this book as well.

The fast stuff is sometimes expressed in *kilobits per second* (1000 bits per second). The abbreviation is *kbit/s*, but you'll often see KBPS, Kbps, and kbps. They all mean the same thing. (You'll also see kilobaud, Kbaud, and kbaud, but remember: they are incorrect).

PITFALL!

> Your modem includes a DAA (data access arrangement) circuit. Years ago, this circuit (which protects the phone company's equipment from undesirable electrical signals) had to be obtained as a separate item from the phone company. Without its inclusion, modems cannot get approval from the FCC (Federal Communications Commission) or the FCC's Canadian counterpart. Other countries require equivalent (but not identical) circuits, and the modem you buy here is unlikely to be approved for operation outside the U.S. and Canada. You should bear this in mind if you're planning to move overseas.

Ready for the Next Step?

Now that you have some of the fundamentals, you're ready for some of the specific things you will have to look for when you purchase a modem. In this chapter, you learned:

- A modem is a device that allows computers to communicate over ordinary telephone lines.
- The telephone was not originally designed for data communication.
- Data is presented at the COM port in the form of negative and positive voltages.
- A modem converts the signal at the COM port to (and from) signals that can be carried over phone lines, such as tones.
- "Baud" and "bits per second" do not mean the same thing.
- Bit/s, bps, and BPS all mean "bits per second."
- Internal and external modems perform the same function.
- It's useful to know what all the connectors look like.
- The connection between the COM port and the modem is called the DTE-DCE interface, and the EIA-232-D standard defines the circuits it uses, as well as their electrical characteristics.

CHAPTER 3

Buying the Right Modem for You

Chapter 2 explained quite a bit about what modems are and how they work. But that's only part of what you need to know. If you're to buy the right modem, you'll have to know and understand the terms used to describe them.

First, though, what are you getting into? Let's look at prices.

How Much Do Modems Cost?

You can buy a modem for under $100. If you have money to burn, you can also buy one for over $1000. The price range for the kind of modem you would expect to use at home with your personal computer, however, is from under $100 to over $400. Look at a very simple table, 3.1, for a summary of pricing information.

Table 3.1 Modem price ranges, by operating speed.

Maximum Modem Speed (bps)	*Price Range*	*Comment*
300	?	Obsolete
1200	$60–$120	Forget it!
2400	$80–$150	Marginal
9600	$180–$400	Best right now
14400	$200–$500	The coming thing

The price ranges in Table 3.1 are approximate, and may have come down dramatically (especially at the high end) by the time you read this.

The comments against the 300- and 1200-bps modems aren't really intended to be funny. 300-bps modems were typically used with dumb terminals for very undemanding interactive applications (not much data went in either direction). 1200-bps modems were used with personal computers for a while—just as long as 2400-bps modems were still relatively expensive.

WORD POWER

The original *dumb terminals* were also known as "glass teletypes." Although they had a display screen rather than a printing mechanism, they offered very little functional improvement over the old teletypewriters.

Why are 2400-bps modems marginal now? First, the prices for the faster modems are coming down rapidly; second, faster modems give you a lot more for your money. You'll learn why this is true a little later in this chapter. In the meantime, let's look at some other burning questions.

COMMON SENSE

Never pay *MSRP* (manufacturer's suggested retail price) for a modem. Shop around. Competition is strong, and the prices are falling all the time.

Why Do Some Modems Have Speakers?

Older modems couldn't dial your call for you. The phone was plugged into the modem, and the modem into the wall socket. To make a call to a remote computer, you would switch the modem to "talk mode," dial the number, listen for an answering tone, then switch the modem to "data mode" and hang up the phone.

Some modems still let you do that, if you really want to. However, when they dial the number for you, the best way to provide you with the warm feeling that things are going according to plan is to let you hear what's going on.

With a modem speaker, you can hear the dial tone, the dialing process, the answer tone, a rushing sound, then silence. From that moment on, the remote computer, your software, and your computer's display screen will keep you informed.

A speaker isn't essential, but it's reassuring. It doesn't usually cost extra, and you don't absolutely have to use it. (See Appendix B if you want to know how to turn it on and off.)

> Unless you can get a real deal on a modem without a speaker, get onc with a speaker.
>
> **BEST BET**

Should I Get an Internal or an External Modem?

You'll have to get one or the other. There isn't a third kind.

Seriously, though, if you have only one personal computer and no plans to acquire a second one, an internal modem is a good bet. It doesn't need to be separately plugged into a power socket and doesn't require that extra cable (called, not surprisingly, a *modem cable*). It may also be marginally cheaper than the equivalent external modem, although the price difference is usually small enough to not be an issue.

An external modem gives you more flexibility. If you have more than one computer, and might not always use the same one for data communications, it's nice to be able simply to move the cable from one computer to the other. Your computer store can even sell you a switch to allow you to have the modem permanently cabled to two or more computers.

Remember, though, it's easier to spill your coffee on an external modem.

Half- or Full-Duplex—Is This an Issue?

Half-duplex operation allows you to transmit or receive, but not both at the same time. *Full-duplex* operation allows simultaneous transmission and reception. Figure 3.1 illustrates the difference between half- and full-duplex operation.

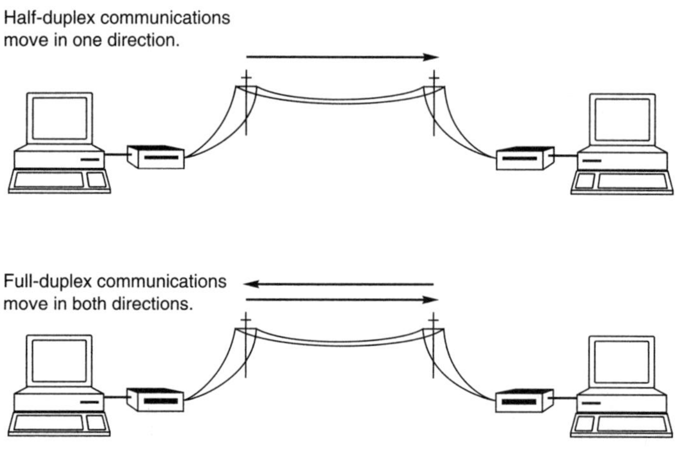

Figure 3.1
Some modems communicate one way at a time. Most of today's modems allow concurrent communication in both directions.

Surprisingly, this isn't an issue. In the bad old days, only the slowest modems would support full-duplex operation over a dial-up connection. The faster ones required a leased line, which was OK if communication was always with the same place.

Today's modems are smarter. Two of them can, in effect, talk at the same time, with each of them filtering out its own signal and listening only to the other. (In case you're interested, this is called *echo cancellation*.) As a result, half-duplex modems are now almost nonexistent. Most communication software can be set up for full- or half-duplex operation. With today's modems, however, it's hard to imagine a situation that would require you to specify half-duplex.

PITFALL!

> If you're considering buying a used modem, *beware*! If it's an older one, not only might it not be full-duplex, it also may not be compatible with any of today's online services or bulletin board systems.

What Is Hayes Compatibility?

Some years ago, Hayes Microcomputer Products, Inc. introduced a range of modems which could respond to a series of commands from the computer (or other device) to which they were attached. Hayes defined commands for modem setup, dialing, running diagnostics, and so on. They also defined responses to the commands, so that the modem could provide the computer with confirmation that a command was accepted, or with progress or status information.

The commands, now known as the *Hayes AT Command Language*, have become a de facto industry standard. If a modem is advertised as being "Hayes-compatible," it means that it supports some or all of the Hayes AT Command Language. At the very least, it means that it supports the commands used for automatic dialing and for switching the modem between command mode (when it can recognize commands) and normal mode (for transmitting and receiving data).

BEST BET

> Don't consider a modem that isn't Hayes-compatible.

What Do the Ads Mean?

Figure 3.2 shows the text from a typical modem ad.

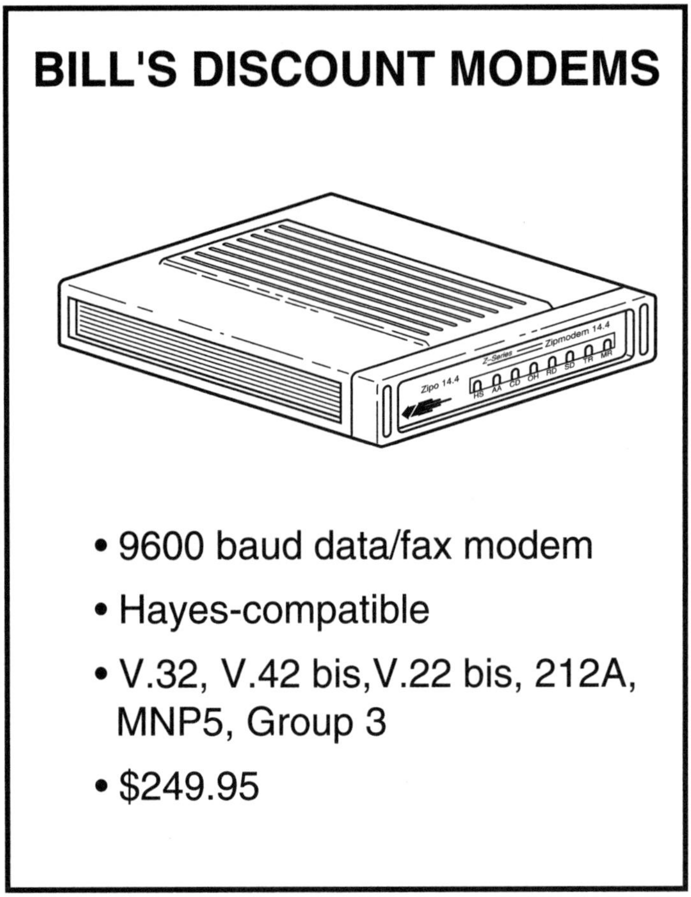

Figure 3.2
A typical modem advertisement.

Let's look at it item by item. (Full explanations of terms can be found in the Glossary.)

- "9600 baud" means that the modem operates at a speed of up to 9600 bits per second. (From Chapter 2, you know that "baud" isn't really the right word, but you've got the advertiser's message.)

- "Data/fax" indicates that it can be used for data transmission (as you'd expect) and, with PC-based fax software, for faxes. Data/fax modems don't look any different from other modems, by the way. They simply have the capability of communicating with fax machines (or, obviously, other data/fax modems).

- "Hayes compatibility" tells you that the modem will support automatic dialing with just about any communications software package you care to buy.

- "V.32" confirms that the modem really does transmit and receive data at 9600 bps. Some data/fax modems can send (and sometimes receive) faxes at 9600 bps, while handling data at only 2400 bps. V.32 modems can also operate, when required, at 4800 bps.

- "V.22 *bis*" indicates that the modem can also operate at either 1200 or 2400 bps. V.22 *bis* is an international standard.

WORD POWER

CCITT uses *bis* (from the Latin word for "twice") to denote either a second version or an interim version of a recommendation; *ter* is used for a third version.

- "212A" also indicates 1200-bps capability. However, 212A (really Bell 212A) is a U.S. and Canadian standard.

- "V.42 *bis*" tells you that the modem has both data compression and error control capability (both of which are discussed later in this chapter). V.42 *bis*, which is the data compression standard, requires that the V.42 capability (error control) also be present.

- "MNP5" is an alternative standard for data compression. It includes the MNP4 error control standard. (More about MNP in the next section.)
- "Group 3" indicates that, when used for faxing (with appropriate PC software), the modem can communicate with Group 3 fax machines (virtually any fax machine produced in the last few years).
- The price is fairly reasonable (for early to mid-1993), by the way—but you can probably do better than $249.95.

This particular set of capabilities is fairly standard in today's 9600-bps modems. Having V.32, V.22 *bis,* and Bell 212A capabilities means that it can communicate with a variety of other modems, up to and including ones with the same set of features. V.22 *bis* provides for operation at 2400 bps, and at the V.22 rate of 1200 bps. Bell 212A provides for operation at 1200 bps in a manner compatible with most existing U.S. and Canadian 1200-bps modems.

For data compression and error control, some modems have either V.42 *bis* or MNP5 capability, but not both. This modem can give you the benefits of data compression no matter which of the two standards is supported at the other end of the line.

Many modem advertisements are much less clear than the one we just examined. Look at the ad from a mail-order supplier in Figure 3.3, for example.

The logo at the top tells you that these are all Intel modems. The first two, at 2400 bps, are almost certainly V.22 *bis* (the accepted 2400-bps standard, which includes V.22 capability at 1200 bps). Probably they also support Bell 212A operation at 1200 bps, for compatibility with all those older modems. But ask—don't assume.

The third is presumably internal (the "B" is a clue, but not a guarantee). Otherwise it and the fourth one seem to have identical capability—which most likely would be that of the first two modems, plus some error control, and possibly data compression. "MNP" could mean MNP3, MNP4, MNP5, or MNP7. Again, you have to ask. These days you should not accept less than MNP5.

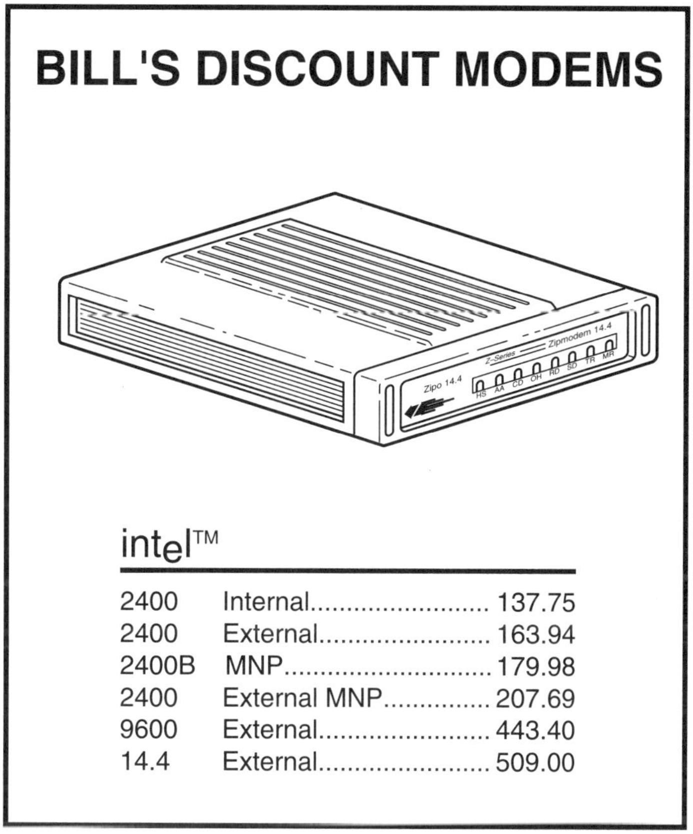

Figure 3.3
Another modem advertisement.

The "9600 External" (9600 bps) probably has all the capabilities of the modem in our first example. How do you know that? The indicators are the price (fairly high), and the fact that Intel entered the modem business fairly recently. Although these factors make this a fairly safe bet, you need to check.

"14.4" is shorthand for 14,400 bps (also written as 14.4 kbps). This modem will have all the capabilities of the 9600-bps modem, plus V.32 *bis* capability. V.32 *bis* is the accepted standard for operation at 14,400 bps.

When mail-order suppliers place advertisements like this one (which is actually a small part of a much larger advertisement), they expect prospective buyers to call and ask questions.

PITFALL! Never buy a modem from a mail-order supplier without finding out its full specifications. And remember, you don't have to place the order on the first call. Shop around!

Would you buy any of the modems in these ads? Let's look a little more closely at all the terms, and see what is the best modem for you.

Bell, Hayes, Microcom, and V-Something

Some of the older modem standards were set by the Bell Telephone Company (prior to the breakup into AT&T and the Regional Bell Operating Companies).

Hayes Microcomputer Products created the widely-used automatic dialing standard you learned about a few pages ago.

Microcom, Inc. developed the Microcom Networking Protocol (MNP) for *error control* and *data compression*. You'll learn more about these shortly. MNP is broken into nine classes—MNP Class 1 to MNP Class 9. MNP Classes 1 to 4 deal with error control. MNP Classes 5 to 9 deal with data compression and other techniques for improving performance.

WORD POWER

There is also an MNP Class 10 standard, sometimes referred to as *fall-back/fall-forward capability*. It defines a protocol for switching to lower-speed operation if the line is noisy, and back to higher-speed operation if the line quality improves. It depends, of course, on the modem at the other end having the same capability.

WORD POWER

A *protocol* is an agreement between parties on how they will exchange information. In this case, the "parties" are modems.

"V-something" refers to the V-Series Recommendations of CCITT (which stands for Consultative Committee International Telegraph and Telephone). As part of the International Telecommunications Union (ITU), based in Geneva, Switzerland, CCITT makes recommendations. Other organizations—such as ISO (International Organization for Standardization) and ANSI (American National Standards Institute)—adopt these recommendations as standards.

Recommendations V.10 to V.33 deal with interfaces and *voice-band* modems (those that communicate over ordinary telephone lines). You need to know about only four or five of them.

Basic Capabilities

The most basic modem characteristic is its *speed of operation*. You need to be concerned about speed in any of the following cases:

- If you will be using anything other than your phone company's no-charge local services for your data communications calls.
- If you will be using an online service that charges for connect time.
- If you expect to upload or download large files (see Chapter 15, "Transferring Files").

Operation at 1200, 2400, 4800, 9600, and 14,400 bps is covered by several standards. Although Bell 212A and CCITT V.22 both support operation at 1200 bps, they are not compatible with one another.

- CCITT recommendation V.22 *bis* supports both 2400- and 1200-bps operation, and is compatible with V.22 at 1200 bps.

- CCITT recommendation V.32 supports 9600- and 4800-bps operation.

- CCITT recommendation V.32 *bis* supports operation at 14,400, 12,000, 9600, 7200, and 4800 bps. At speeds of 9600 and 4800 bps, it's compatible with V.32.

All standards defining the speed of operation also define a number of other characteristics, including the techniques to be used for modulation and demodulation. The speed is not an absolute measure of performance. Error control and data compression also help.

Error Control and Correction

Errors on a communication line can slow you down. Error-correcting modems adapt in a number of ways to certain kinds of line problems, and can even correct some kinds of errors on the fly (without having to retransmit).

CCITT V.42 and MNP Classes 2, 3, and 4 are error-control standards. Class 4 includes Class 3, and Class 3 includes Class 2.

Data Compression

Your modem can send a more efficient signal by compressing data. Although there are some variations on this theme, the two main ways for a modem to compress data are:

- By *translating* it into a code that uses fewer bits for the characters that occur most frequently.

- By replacing repetitions of the same character with a special code, followed by one occurrence of the character, and a count (which tells the receiving modem how many occurrences of the character it must re-create).

V.42 *bis* and MNP Classes 5 to 9 are data compression standards. The most common are V.42 *bis* and MNP Class 5.

> As V.42 *bis* requires V.42, and as MNP Class 5 includes MNP Class 4, you can assume V.42 when you see V.42 *bis,* and you can assume MNP4 when you see MNP5 in a modem ad.
>
> **COMMON SENSE**

What Does It All Boil Down To?

Now you know all the code words, you can read and understand the ads.

In making your selection, remember that high speed, multiple speeds, error control, and data compression are not luxuries. They all increase your flexibility with respect to the online services you can use. Some services support error control, data compression, and optional support for higher-speed operation; if you have an appropriate modem, you can take greater advantage of these capabilities.

Operating Costs at 2400 and 9600 bps

There have been several times of transition in the personal modem business—from 300 to 1200 bps (early 1980s), from 1200 to 2400 bps (late 1980s), and now from 2400 to 9600 (and even 14,400) bps. Today's 9600-bps modems are available at last year's 2400-bps prices. Also, this year's 2400-bps modems are available at last year's 1200-bps modem prices. But don't forget: the modem is not your only cost—you may have phone charges (or charges for being online) as well.

Let's say you want to download some shareware from CompuServe. The file is about 450,000 bytes in length. For such a download, CompuServe charges $8/hour for 1200- or 2400-bps service, and $16/hour for 9600-bps service. Which service is more costly? The answer may surprise you. Table 3.2 shows the approximate elapsed times, and the costs incurred.

Table 3.2 CompuServe File Transfer Cost Comparison

Speed (bps)	*Time Rate/Hour*	*(min)*	*Cost*
1200	$8	70	$9.33
2400	$8	35	$4.67
9600	$16	9	$2.40

If you don't have a local CompuServe access number, you'll have to add the cost of the long distance call to these figures.

With data compression, the cost at 9600 bps would be even lower. However, as most of the available files are already compressed (with PKZIP, for example), the opportunity for further compression may be limited (but not totally absent).

COMMON SENSE

> Look at **all** the costs when you choose a modem, not just the modem's price.

Try It!

Following is a hypothetical ad for a modem (Figure 3.4). Read through it, and see if you can now decipher it.

See if you can answer these questions.

1. What is the maximum data transmission speed of this modem?
2. Does it support automatic dialing?
3. If I use it to communicate at 1200 bps, does it matter what kind of modem is at the other end of the line?

4. Does it have data-compression capability?

5. Does it have error-control capability?

6. Can it send *and* receive faxes?

7. Is it a good buy?

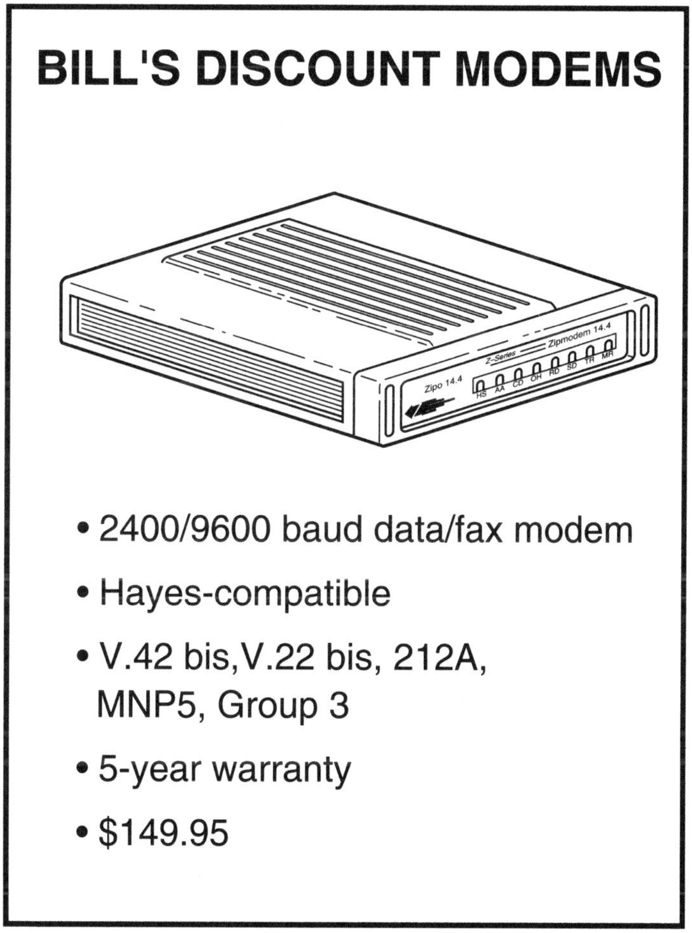

Figure 3.4
Hypothetical ad for a modem.

Now, check your answers.

1. 2400 bps. The fastest standard shown is V.22 *bis*. Also, 2400/9600 in conjunction with data/fax indicates that the 9600-bps rate is good only for faxing.

2. Yes. That's what Hayes-compatible means.

3. Probably not. It supports both Bell 212A and V.22 standards at 1200 bps (V.22 *bis* includes V.22). Although there used to be other standards at this speed, they're rare enough today that it's a good bet the modem you reach will support one of these two.

4. Yes. Both V.42 *bis* and MNP Class 5 are data-compression standards. As long as the modem at the other end of the line supports one of them, yours can take advantage of data compression.

5. Yes. The V.42 *bis* standard *requires* V.42, and MNP Class 5 *includes* MNP Class 4. Both of these are error-control standards.

6. Not necessarily. In fact, it's unlikely that a 2400-bps modem will support receiving faxes (which may not be a big loss—see Chapter 5). 9600-bps modems with fax capability usually support both sending and receiving.

7. No. For about $40 or $50 more, you can get a V.32 (9600-bps) modem, with error control, data compression, and fax capabilities. The 5-year warranty is attractive, but with the reliability of today's integrated circuits, the probability that you'll need it is very low. If the modem is defective, you'll know within the first year.

COMMON SENSE

> You'll want a better, faster modem in a year or two. Don't pay more just to get a longer warranty.

At the end of Appendix D, you'll find modem purchase checklists. These will allow you to organize and compare the essential characteristics of a number of competing modems.

Ready for the Next Step?

No matter how good your modem is, you won't be able to use it without software. Chapter 4 provides some tips and a number of examples. In the meantime, you have now added the following to your grab-bag of modem knowledge.

- You can expect to pay $180-$400 for a full-featured modem.
- A modem speaker lets you hear what your modem is doing, such as dialing or connecting.
- You should choose an internal modem if you'll use it with one computer, and won't move it much; choose an external modem if you'll be using it with more than one computer, or moving it around.
- You need to get a Hayes-compatible modem, which supports such features as automatic dialing.
- You're only interested in full-duplex modems; half-duplex modems are obsolete; they use up too much time and money.
- Modems conform to a number of standards.
- Some standards (Bell 212A and CCITT V.22, V.22 *bis,* V.32, and V.32 *bis*) address the speed of operation and the method used for modulation and demodulation.
- Some standards (CCITT V.42 and V.42 *bis,* and the MNP series) address error reduction, error correction, and data compression.
- The Hayes AT Command Language handles automatic calling and answering.
- As long as two communicating modems have at least one modulation standard in common, they can communicate.

- When two modems communicate, if one has more features than the other, the excess features won't be used (but they won't hurt, either).
- You ought to buy the fastest modem you can reasonably afford.
- Error control and data compression are valuable features.

CHAPTER 4

What About the Software?

To use your personal computer for communication, you need to make it operate like a terminal—or better! Your computer only becomes the DTE (data terminal equipment) you learned about in Chapter 2 when it is running *terminal emulation* or other *communications software*.

> **WORD POWER**
> *DTE (Data Terminal Equipment)* means a device at which data transmission originates and/or terminates. This may be an actual terminal, or a computer running communications software.

There are dozens of communications software packages on the market today. Some are *generic*, and can be used to communicate with many of the online services and bulletin boards, or with the personal computer of a friend or colleague. Others are specific to a particular online service, and can be purchased in one of these ways:

- From the provider of the service (such as CompuServe).
- As part of a startup package you can buy at a computer store or (in some cases) a bookstore.

Deciding What You Need

Are you going to focus on a single major online service? Are you going to subscribe to two—or possibly three—major online services? Are you going to explore the local bulletin board systems? Are you going to communicate directly with a friend or colleague who also has a modem?

Depending on how you answer these questions, the scope of your search may be limited or quite broad.

For example, if you subscribe to PRODIGY, you have no choice. The software they provide with the membership kit is the only software that works.

America Online gives you a choice of two software packages, depending on whether or not you use Windows. Again, however, you must use one of the two. Nothing else will work.

CompuServe is somewhat more flexible. You can choose from several software products that are specific to CompuServe, or you can use generic software. If you intend to be a heavy user of CompuServe, however, you'll be much better off with one of the CompuServe-specific packages.

We discuss the PRODIGY, CompuServe, and America Online software later in this chapter. In the meantime, given that these three don't occupy the entire universe, let's consider the communications software features you might need for other online services, for bulletin board systems, and for electronic schmoozing with remote colleagues. First, we'll throw some features at you, then we'll discuss each of them.

Here are the features:

- Terminal emulation
- File transfer support
- Script language and "canned scripts"

- Security support
- Specific modem support
- Phone and address books
- Help screens
- Text editor
- True high-speed support

Terminal Emulation

WORD POWER

Terminal emulation refers to the use of your computer as if it were a terminal. Through the magic of software, your computer can be made to mimic any of a large variety of terminal types.

There are three very basic terminal emulations, all of which are supported by most communications software. These are:

- TTY
- DEC VT52
- DEC VT100

TTY emulation turns your PC (or, for example, a window within Microsoft Windows) into a Teletype terminal. What appears on your screen is what would be printed on the continuous roll of paper on a real Teletype machine. The characters you can enter and display are limited to the 94 letters, numerals, punctuation, and special characters of 7-bit ASCII.

WORD POWER

ASCII stands for the *American Standard Code for Information Interchange*. ASCII characters are made from a standard data communication code, in which each character is made up of seven bits. This provides for a maximum of 128 unique characters. The first 32 characters (values 0 to 31) are reserved as "control characters." Of the remaining 96, the first (value 32) is a space and the last (value 127) is an idle character. This leaves 94 values to represent the characters that can be displayed or printed. Personal computers use an 8-bit extension of ASCII, providing for up to 256 unique characters.

TTY emulation is acceptable for certain online services, including CompuServe, but it's about as basic as you can get.

Emulation of Digital Equipment Corporation's (DEC's) VT52 and VT100 terminals will give you much more flexibility. It's useful if you want to dial into some of the neighborhood bulletin boards. If you're using a Windows-based product like Windows Terminal, try VT100. If the online service or neighborhood bulletin board supports VT100 emulation, it can liven up your display (and make it more legible) with characters that are bold, underlined, double width, double height, and so on.

Most communications software products can emulate a wide range of terminal types. All of them include VT100 emulation; some also support later DEC terminals, such as the VT220 and VT240.

File Transfer Support

WORD POWER

File transfer is the copying of files from a device at one end of a communications link to a device at the other end of the link. File transfer from an online service is usually called *downloading*. File transfer to an online service is usually called *uploading*.

File transfer capability is essential if you are to take advantage of the various files you can download from local bulletin board systems and other online services. The contents of these files may range from recipes to short stories, from cartoons to naughty pictures, from programs to documentation, from political manifestos to religious tracts, and so on. Just as there are modem protocols (see Chapter 3), there are protocols for the transfer of files. A good communications software package supports at least the most common protocols (which we describe in Chapter 15).

Script Language and Canned Scripts

A *script language* allows you to create procedures for automating some of your online tasks. The very least of such tasks is that of making the connection to an online service and signing on. Beyond that, you can automate the retrieval of E-mail, your navigation through various sections of the online service or bulletin board system, and so on.

Some communications software packages provide "canned" scripts to help you sign on to and move around within some of the better-known online services, such as Delphi, GEnie, Lexis/Nexis, and so on.

Some packages have a "learn by example" feature. Once you are familiar with some of the things you have to do on your favorite online service, such as obtaining a stock quote, you can turn on the software's *learn mode* and allow it to "remember" your keystrokes, the responses you receive, and the choices you make. You can then turn off the learn mode, and give a name to the script it has just generated. The next time you want to get the stock quote, you can ask the software to run the named script. For the beginner, this learning feature avoids the need to learn how to write scripts—something you might want to do as you gain experience.

A script capability is good. Canned scripts and a learn mode make that capability very useful. What's even more useful is a script language that is popular, which means that there will be users uploading useful scripts (with descriptions) into one of the online service's libraries, from which you can download the ones you think you can use.

Security Support

Some bulletin board systems and online services have security features to limit access. Some require that you dial their number, submit a code, hang up, and respond when they call you back. In Appendix B, we discuss how this can be handled automatically in the Hayes AT Command Language. With security support, a communications software package can create and use the appropriate AT command strings for you. When you sign up for a secure online service, you provide them with the phone number of your modem, and they provide you with a security code. Then you provide your communications software package with a name for the online service, their modem access number, and the security code. When you want to sign on, you select the service by name, and the software does the rest.

(If someone steals your security code, there's no problem. The online service will call your number, not the thief's.)

Specific Modem Support

When you configure a communications software package, it will give you a list of modem types from which to select yours (by make and model). It will also provide you with some generic descriptions, such as Hayes-compatible. Although Hayes compatibility means it's not essential that your modem have a specific entry in the list, its inclusion will allow the software to take full advantage of its specific capabilities. This saves you from having to pore over your modem manual (though it's still a good idea to read it).

As an alternative to providing you with a specific modem list, your software may test your modem with a series of Hayes AT commands at some point in the original procedure of installation and setup. The modem's responses (including error indications for commands it doesn't support) will allow the software to determine its capabilities.

Phone and Address Books

You can use a communications software package to dial an online service or bulletin board system. With some packages, you can also dial voice calls.

In either case, your life will be easier if the phone number can be easily retrieved from the electronic equivalent of a card-index file.

In addition, you may want to use one or more online services for E-mail. Again, you need an electronic "address book" so you can avoid having to enter E-mail addresses (some of which might be quite long) every time.

All the popular communications software packages support phone and address books.

Help Screens

A *help function* is very useful. It saves you from having to turn to the manual all the time. Even better is *context-sensitive help*—which means that when you ask for help (by pressing F1 or some other specified key), the information displayed will be relevant to where you are at the time.

Almost all the popular communications software packages have context-sensitive help screens.

Text Editor

When you are composing E-mail messages, there are few things more annoying than the inability to edit the text, especially if you're unable to rearrange it with a cut-and-paste function. Luckily, the popular communications software packages either include a *text editor* program, or provide a means for you to incorporate an existing editor (such as MS-DOS EDIT).

True High-Speed Support

If your modem has MNP Class 5 or V.42 *bis* data compression, it can accept data from your COM port faster than its rated transmission speed. For example, a V.32 *bis* modem, which operates at 14,400 bits per second, might be able to achieve a compression ratio of 4 to 1 if it has V.42 *bis* data compression. To take advantage of this, your computer's COM port speed would have to be 57,600 bits per second. This means that you must be able to specify this speed to your communications software package.

In fact, if your computer doesn't have the latest communications chips (16550 UARTs—see *UART* in the Glossary), its communications performance will still fall quite a bit short of 57,600 bits per second (but still do better than 14,400). If speed is really important to you, and your computer doesn't have the new UARTs, check with one of your local computer-store gurus to see if an upgrade is possible; if it is, make sure you get a communications software package with specific 16550 support.

Popular Communications Software

Now let's look at some specific software packages.

Windows Terminal

If you are a user of Microsoft Windows 3.0 or 3.1, you already have the software to run your PC as a terminal. When you installed Windows, the Setup program placed the Terminal icon in the Accessories Group. (You may, of course have subsequently moved it to another group.) Windows Terminal is actually a generic communications program.

When you start Terminal, you may set it up to emulate one of three terminal types, to conform to the modem and port through which you will communicate, to dial a given phone number on request, and so on. Then you can pull down the **File** menu and use the **Save As** command to save your configuration to an appropriate file name (such as **mcimail.trm**). In version 3.1 of Windows, you can then create an icon for your configuration by dragging the file name from the Windows File Manager to the Accessories (or other) Group. Finally, using the **Properties** option of the Program Manager's **File** menu, you can enter an appropriate description (**MCIMail**, for example).

BEST BET

Configure and save a different version of Terminal for each of the online services with which you intend to communicate.

> If you are already a Windows user, use Windows Terminal to get started (except on PRODIGY or America Online, where you have no choice). Take your time selecting software for longer term use. (That's two Best Bet Tips in a row. How lucky can you get?)
>
> **BEST BET**

Figure 4.1 shows the Settings pull-down menu of Terminal. Chapter 9, "Getting It Running," illustrates how you can configure some of the attributes that are listed.

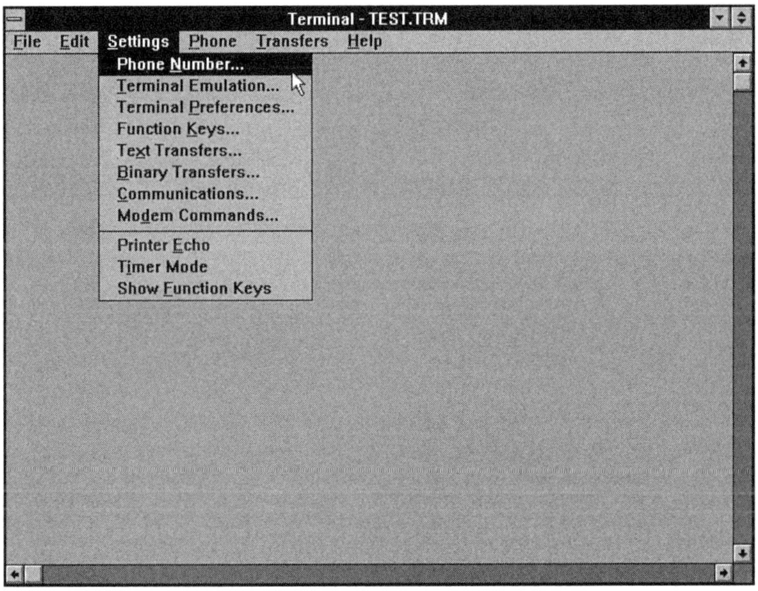

Figure 4.1
Windows Terminal screen with settings menu.

Other Windows-Based Software

The June 30, 1992 issue of *PC Magazine* reviewed 23 communications software products, including several for Windows. Their favorite Windows product was DCA's Crosstalk for Windows, version 1.2.1. It has a list price of $195. Here and there in this book, we admonish you to "always read the manual," but according to *PC Magazine*, installing Crosstalk is so easy that you don't need the manual (but read it anyway). Crosstalk for Windows provides specific support for five major online services—CompuServe, Dow Jones News Retrieval, Lexis/Nexis, MCIMail, and OAG (Official Airline Guides). A later version is now available (Crosstalk for Windows 3.1). Calling and signing on to CompuServe, for example, is just a matter of pointing to an icon and clicking your left mouse button twice.

PC Magazine also gave an honorable mention to Software Ventures' Microphone II for Windows, noting that it was very fast, but had somewhat fewer features than DCA's product. At $129, however, its list price is also somewhat lower, and the reviewers considered it easy to use. It does include specific support for CompuServe, Dow Jones, and MCIMail.

Although it couldn't provide a full review, *PC Magazine* gave kudos to the about-to-be-released ProCOMM Plus for Windows, a $149 package which provides extremely high performance with its support of the 16550 UART (see *UART* in the Glossary).

BEST BET

> Look for specific support ("scripts") for the online services you intend to use. It will make life easier. The products designed specifically for a service do, of course, make life (when using that service) easier still.

The number of Windows-based packages increases all the time. New versions of existing packages continue to replace old versions. Shop around, read the magazines and the weekly computing section of your local paper (if it has one), and get feedback from any of your friends who are already online freaks. Do the same for the DOS-based packages, too.

QMODEM

A shareware version of QMODEM, a product of Mustang Software, Inc., is provided at no charge with many modems (around 35, at last count). It is a DOS application, with a wide range of capabilities that users can customize. You must be prepared to be patient at first—there is a vast number of choices, and you'll need to memorize key combinations to activate many of them (unless you want to bring up a Help screen every time). If you like to show off, QMODEM provides plenty of opportunity. Figure 4.2 shows one of QMODEM's installation menus—the Modem Setup menu.

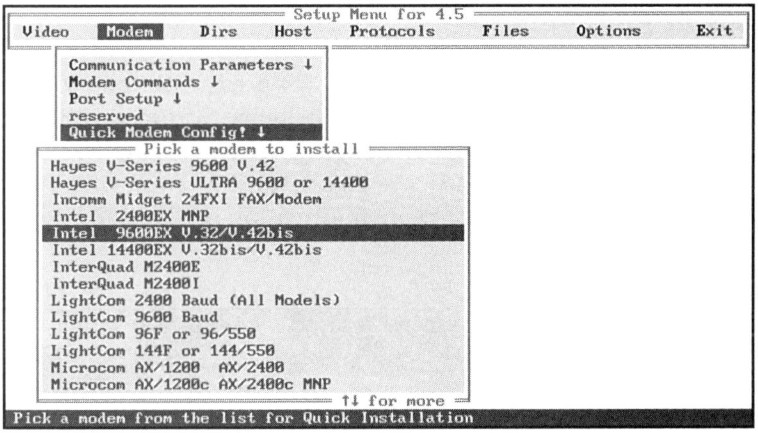

Figure 4.2
QMODEM Modem Setup menu.

If you like QMODEM, you might want to consider Mustang Software's upgrade offer to the "Full Commercial Version" (rated "easy to install" by *PC Magazine*). Mustang is also developing a Windows-based product, and recently started shipping a completely rewritten QMODEM Pro.

Other DOS-Based Software

If you don't use Windows (or are willing to do without it occasionally), you have a wide choice. *PC Magazine's* review was dominated by DOS-based products. Their highest praise went to Hilgraeve's HyperAccess/5, which has a suggested retail price of $99.95. You might want to consider its ease of installation, ease of use, and the specific support of 165 different modem models.

Service-Specific Software

Two major online services require that you use their software—there is no other way to communicate with them. For others, such as CompuServe and MCIMail, service-specific software is offered, and is advantageous but not essential.

PRODIGY

The only way you can use the Prodigy Service is with their software, which requires that your PC have graphics capability. Fortunately, the software supports a range of PC graphics adapters, from the older Hercules monochrome cards to EGAs and VGAs. Figure 4.3 shows you what PRODIGY's initial screen looks like. For more information on PRODIGY, see Chapter 10, "Exploring PRODIGY."

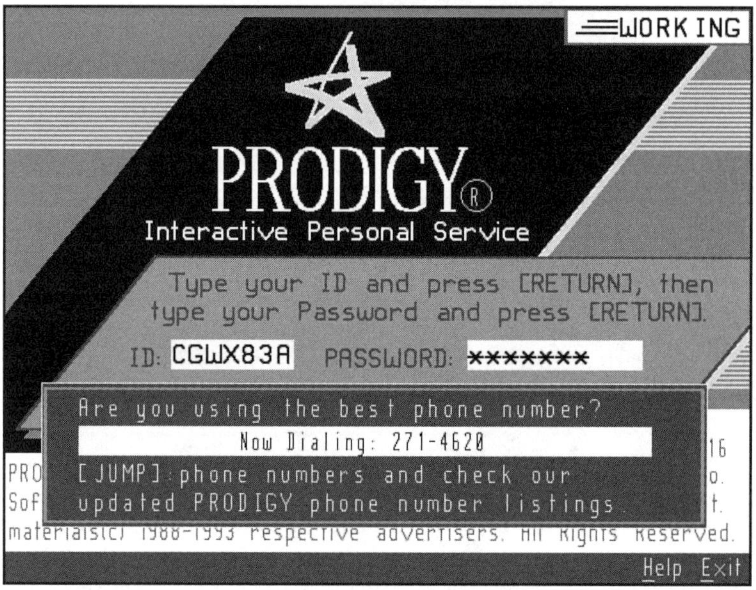

Figure 4.3
PRODIGY's initial screen.

CompuServe

CompuServe comes ready to confuse you with multiple choices. You can dial into CompuServe using any of a number of generic software products. But, you can also choose to get one of several programs that offer easy-to-use menus which help you navigate through CompuServe. (You can learn more about CompuServe, as well as the other programs described in this section, in Chapter 11.)

The product offered and supported by CompuServe itself is CIM, or the CompuServe Information Manager. It's available in both DOS and Windows 3.1 versions.

Figure 4.4 shows the initial screen displayed by CIM Version 2.0. If you choose any of the items on the Browse menu, CIM will dial the local CompuServe number (which you will have supplied at installation time), sign you on, and connect you to the service you have chosen.

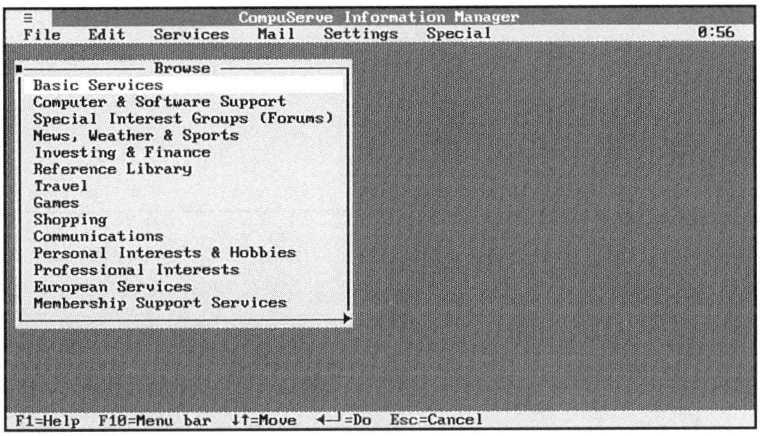

Figure 4.4
Initial CIM screen.

OzCIS is a *freeware* program. It can be downloaded from one of the libraries of CompuServe's IBM Communications Forum (IBMCOMM).

WORD POWER

Freeware is software offered at no charge, and with no expectation of payment. *Shareware*, on the other hand, is software for which there is no initial charge. Users may try it, discard it, and owe nothing. If they continue to use it, however, they are expected (and honor-bound) to pay a fee, at which time they will become *registered users*. Registered shareware users may be entitled to

continues

continued

support, and may even receive an upgraded version of the software. If you download shareware and continue to use the downloaded copy, you may find that every time you use it, it will prick your conscience with a message reminding you that you can avoid annoying messages by paying the license fee and getting the upgraded version.

Although it takes slightly longer to learn than CIM, OzCIS is extremely powerful, and offers some cost-saving features. Signing on, sending your replies, and signing off again is automated. Figure 4.5 gives you a look at OzCIS.

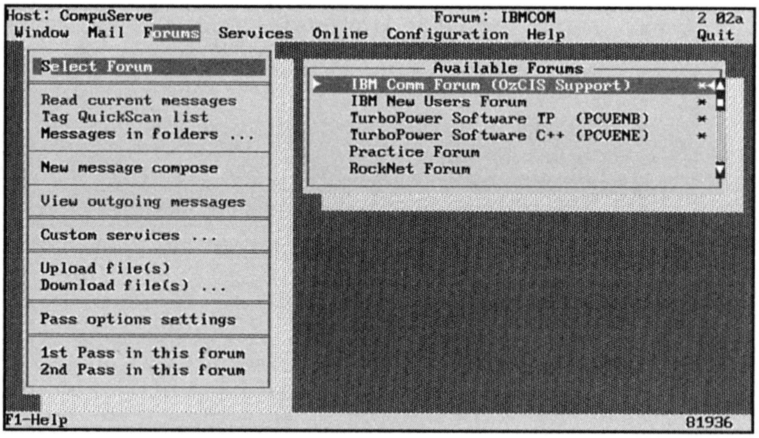

Figure 4.5
Initial OzCIS screen.

The other popular program that is specific to CompuServe is TAPCIS. Its users are very enthusiastic about its capabilities.

BEST BET

If you're going to subscribe to CompuServe (and are a beginning user), buy the CompuServe kit at a computer store. Once you've established your account as described in the kit, sign on (experienced users can get along without the kit). Then go to the DOSCIM forum to download the latest copy of CIM. The charge is normally $25 (sometimes it's on sale for $15), but you get a $25 usage credit against your account. Once you're a registered CIM user, you can get future upgrades for a $10 fee, again with an equal usage credit. Even if you switch to OzCIS or TAPCIS later, you will have incurred no net cost for CIM.

America Online

Like PRODIGY, America Online is accessible only with *proprietary* software. You can buy their *Membership Kit and Tour Guide* for either the Windows or the DOS version of the America Online software. Each is in the form of a book (almost 400 pages), with a 3.5-inch diskette inside the back cover. The kit entitles you to ten free hours online. (There is also a book that is simply called *Tour Guide*, which is mailed to members.)

Figure 4.6 shows the screen that is displayed when you start the Windows-based software.

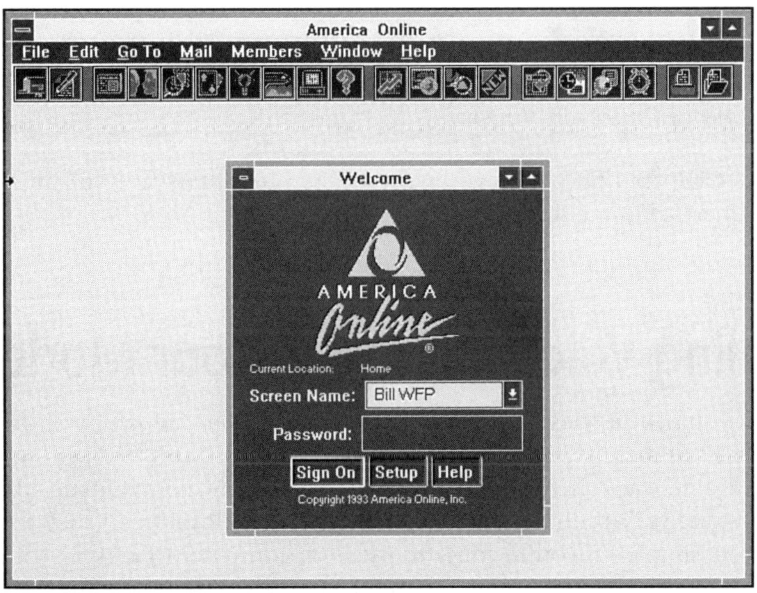

Figure 4.6
America Online for Windows' welcome screen.

Figure 4.7 shows the clear way in which America Online advises you of the progress of your call. The keypad is highlighted during dialing, the connectors are highlighted when the modem connection has been established, and the key is highlighted during password checking.

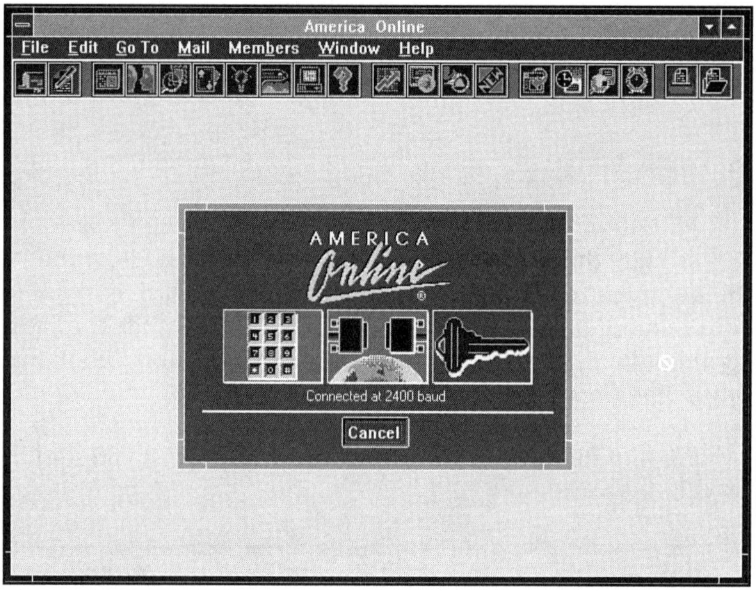

Figure 4.7
America Online for Windows' call progress screen.

Chapter 12 provides you with some examples of how valuable the America Online service can be.

Dialing Voice Calls with Your Modem

If you have a telephone on the same line as your modem, you can use any of a number of *electronic card index* software products to dial your calls automatically. If you have Windows, you can take immediate advantage of Windows Cardfile's dialing capability. The philosophy of such programs is very simple—if you're going to use your computer to look up someone's phone number, you might as well have the computer dial it, too.

Ready for the Next Step?

The next chapter deals with some of the pros and cons of using your computer as a fax machine. Among other things, in this chapter you learned:

- To use your computer for communication, you need communications software—which must provide, at the very least, terminal emulation capability.
- Some online services (PRODIGY and America Online) require that you use their software.
- You can use generic software for CompuServe, but you're better off with one of the CompuServe-specific packages.
- Although limited in scope, Windows Terminal provides a good starting point. You can shop around later.
- The popular communications software packages provide a great deal more than terminal emulation.

CHAPTER 5

The Fax of Life

This is a particularly long and detailed chapter. However, if you have a requirement for faxing, you should read it.

Some modems allow your computer to be used for sending and receiving faxes. This chapter will help you to identify the essential characteristics of such modems and of the software that allows you to use them for faxing. Along the way, it will touch on some of the pros and cons of "faxputing" (a new and eminently forgettable word—you saw it here first).

The High-Priced Fax Machine

If you don't already have a personal computer and modem—and if all you really need or want is a fax machine—forget the computer and forget the modem. Fax machine prices start at not much over $200. Even a combined phone, fax, and answering machine is only a few hundred dollars.

There is one exception to this piece of advice. You may have a business with a number of customers or clients. You may want to send announcements to many or all of them on a regular basis. Moreover, you may want to send such announcements by fax. Some of the more expensive fax machines can be used to *broadcast* a single document (single- or multi-page) to many fax numbers, but entering the list of numbers into the machine's memory (and maintaining the list, once it's entered) can be a real chore. On the other hand, many fax software packages for personal computers include a broadcast capability and the means to set up and maintain broadcast lists easily.

PITFALL!

Broadcasting faxes to your customers and business contacts is one thing. Advertising by fax is quite another. Laws now exist that limit "unsolicited" faxes. They are based on very straightforward logic. When you send a document by fax, the recipient incurs a cost, however minor, which may include fax paper, wear and tear, unavailability of the fax machine (for its owner's own purposes), while your fax is being received, and so on. Consumer protection agencies, a lawyer, or your local Better Business Bureau may be able to advise you on this.

Looking at Fax Modems

If you already have a personal computer and are not interested in any communication capabilities other than sending and receiving faxes, think carefully. You may be able to get a modem with fax capability for $100 or less, and fax software may be included in the price, but it won't necessarily fit your requirements. We'll note some pitfalls, as we go along. In the meantime, let's consider what the modems look like.

WORD POWER

Modems with fax capability are referred to as *fax modems* or *data/fax modems*. They are the same thing. To help you get used to it, we use both terms.

What's the Difference?

You can't go by appearances. Fax modems, both internal and external, look just like any other modem, but they have some additional capabilities. You don't need a detailed knowledge of such capabilities, but you will need a broad understanding of some of the terms that are used.

The first two terms are *Group 2* and *Group 3*. Most of today's fax machines are classed as Group 3, which is the name of an international standard for machines that can transmit a page (letter size) in under one minute (using a transmission speed of 9600 bps). Group 3 machines (and Group 3 data/fax modems) can also communicate with the older Group 2 machines, which take about three minutes to send or receive each page. Group 3 machines have a *fine mode* (better resolution) and a *half-tone mode* (for shades of gray); Group 2 machines don't.

BEST BET

> A new standard—*Group 4*—covers a new class of fax machine, operating at much higher speeds than Group 3 machines (at 19,200 bps over ordinary telephone lines, and at up to 64,000 bps over the higher-priced digital circuits used by "corporate America"). Group 3 machines use data compression, but Group 4 machines do it better, resulting in a much shorter time per page. Group 4 machines must be able to communicate with existing Group 3 machines (who would buy them, otherwise?). You can also expect to see modems with Group 4 fax capability in the near future.

The next two terms are *Class 1* and *Class 2*. They apply, not to fax machines, but to fax modems, and define how the workload is shared between the modem and the fax software on your computer. Class 1 is well-established. The Class 2 standard is new, but is already supported by some data/fax modems and some fax software.

Two more terms are *CAS* (Communications Application Specification) and *SendFax*. CAS was developed by Intel Corporation and enjoys some status as a de facto standard (in other words, other companies have designed their products to work with it, but it has no official recognition). SendFax was developed by Sierra Semiconductor, some time ago, and applies only to sending, not receiving.

One final term is *V.17*. If you were paying attention in Chapter 2, you'll recognize this as a CCITT Recommendation. V.17 specifies Group 3 fax modem operation at 14,400 bps—fifty percent faster than standard Group 3 operation. It's only good for computer-to-computer faxing; the speed will drop to 9,600 bps if a real fax machine is at the other end of the line.

Checklist—Fax Modem Capabilities

Now you know the terms, let's summarize what you have to look for in the modem advertisement or on the box. (Note that all current data/fax modems offer Group 3 capability, plus at least one of the following capabilities: Class 1, Class 2, CAS.)

- Group 4 capability is nice to have, but it might increase the modem's price significantly. It includes Group 3 capability. Software support for Group 4 is limited at the moment.

- Group 3 capability is essential, and all data/fax modems have it. It includes Group 2 capability.

- Class 1 capability enjoys wide support from fax software vendors but is essential only if the modem doesn't have CAS capability, which is also supported by a number of fax software products. If the modem does not have CAS capability, Class 1 support is essential.

- Class 2 capability is very nice to have. Software support for it is increasing. You should see better performance (fast operation) with Class 2—your software will do less of the work. However, be careful; even though the Class 2 standard has only just been adopted (April, 1993), some modem and software vendors jumped the gun and brought out products based on draft versions of the standard. If you use the fax software that's included with the modem it will work (so should the upgrade they're sure to offer you); other fax software may not.

- CAS capability is essential only if the modem does not also have Class 1 capability (see Class 1, above).

- SendFax capability won't hurt. However, if the modem doesn't also have either Class 1, Class 2, or CAS capability, it cannot receive faxes. If you don't want to receive faxes (see "Your Fax Needs," later in this chapter) on your computer, you might be able to get a good deal on a used modem with SendFax capability.

- V.17 capability is nice to have, but, if you have to pay much more for a V.32 bis modem that has V.17 capability than you do for one that does not, its value is doubtful. You'll get no performance improvement when communicating with real fax machines. Note that V.32 bis modems with V.17 capability may be labeled "14400/14400 data/fax modem," and V.32 bis modems without V.17 capability may be labeled "14400/9600 data/fax modem."

In December, 1992, *PC Magazine* reviewed twelve fax modems, ranging in list price from $399 to $1,295 (whew!). All were produced by big-name manufacturers (Computer Peripherals, Hayes Microcomputer Products, Intel, Practical Peripherals, SupraCom, U.S. Robotics, and Zoom Telephonics). Since then, prices for all modems have fallen (at least the actual selling prices have), and most V.32 and V.32 bis modems include fax capability. It's a whole new ball game, and you're going to have to follow the advice you'll find in Chapter 7. Ask questions, including questions about how one fax modem's performance compares with another's.

Checklist—Software the Fax Modem Needs

Fax software is designed to meet two sets of needs—yours and the modem's. The modem's needs are quite straightforward, so let's look at them first.

If your fax modem package includes fax software, the modem's needs will already have been met. If you buy your fax software separately, you'll need to ensure compatibility with your modem. Here's what you need to know.

- If your modem has Group 4 capability, your software should have it too—at least eventually. You might want to forgo it for now and upgrade later. Most fax software vendors can be expected to offer an appropriate upgrade as Group 4 faxes become more common. For now, it might be like the V.17 (14,400-bps) capability—only helpful when faxing to or from other computers.

- If your modem has Class 2 capability in addition to either Class 1 or CAS (or both), look for software that has Class 2 capability, too. Better than three out of four fax software products have it, so demanding it doesn't appreciably limit your choices.

- If your modem has only Class 1 or only CAS capability, make sure your software and modem support the same one. Class 1 gives you twice as many software packages from which to choose. Version 1.1 of Eclipse FAX (from Eclipse Systems, Inc.), one of the fax software packages to which *PC Magazine* gave its Editors' Choice award in December, 1992, was one of only two products combining support for CAS with lack of support for Class 1. Version 1.2 of the same software, released since then, supports Class 1. (An Editors' Choice award was given to one other product—FaxMaster, from Caere Corporation.)

- Make sure your fax software supports fax transmission at all your modem's specified speeds up to its maximum fax speed. For the reason we've already stated (that real fax machines are limited to 9,600 bps), the 14,400-bps capability is not immensely important—but most fax software supports it anyway.

What You Need from the Software

Different fax software products have different strengths—and weaknesses. Relative to using a fax machine, computer-based faxing, in general, has strengths and weaknesses. We're going to look, specifically, at what the software can do for you. In doing so, we'll address the pros and cons of using your computer as a fax machine.

In its December, 1992 evaluation of fax software, *PC Magazine* summarized each product in terms of its suitability for four general tasks: sending, receiving, fax management, and printing. Let's look at these in turn.

Sending

With a fax machine, sending is quite straightforward: insert the document, optionally choose the mode (fine, or half-tone—the default is usually normal), dial the number, wait for the high-pitched tone, and press start. You may not have to actually dial the number; most fax machines have one-touch dialing for a few numbers and speed dialing for quite a few more. The main prerequisite, of course, is that you already have a document (not so with fax software).

In fact, with fax software, sending an existing (hard or paper copy) document is impossible unless you also have a scanner. This leads us to a discussion of how fax software sends faxes.

Whether your faxes are created from text files, word processing files, desktop publishing files, or graphic files, they must be rasterized for transmission, to conform to the standard defined for Group 3 (or Group 4) fax machines. The software handles rasterizing for you without being asked, but, if your computer isn't one of the latest models, it can be time-consuming.

Rasterizing is the process by which your fax software converts a text document into a form that can be transmitted to a fax machine. It is the same process used by your computer's video card to produce the image on the screen, which consists of a series of scan lines, similar to those on a TV screen. The process of "writing" to the screen is the same, whether the image is all text or graphics. Now, take a deep breath. . . .

A fax machine reads a document one scan line at a time, advancing the paper for each scan line. For each inch of each scan line, it generates 203 numbers (in normal or fine mode, they are one-bit numbers—1 for white, 0 for not white; in half-tone mode, they are four-bit numbers, ranging in value from 0 to 15, for sixteen shades of gray). The receiving fax machine steps and scans in the same way, "burning" the image into its sensitized paper according to the value of the received numbers. In normal mode, there are 98 scan lines per inch; in fine or half-tone mode, there are 196. Relax; we're nearly there. . . .

We'll conclude this somewhat lengthy explanation with an example. If you have a document file whose contents would normally fill one sheet of letter-size paper, the rasterizing process will convert it to 1,078 (for normal) or 2,156 (for fine or half-tone) "lines," with 1,725 numbers per line (one-bit numbers for text or line drawings, or four-bit numbers for full graphics). That may seem like an awful lot of bits (from 1,859,550 to 14,876,400), but data-compression techniques reduce their number dramatically.

Let's look at the document possibilities.

- Existing documents.
- Original text (produced with a word processor, text editor, desktop publishing application, and so on).
- Original illustrations (produced with graphics software, such as CorelDRAW!, Adobe Illustrator, and so on).

In order to use your computer to send original hard copy documents, you need a scanner. Like a fax machine, a scanner converts a document to digital form by rasterizing it. The easiest scanners to use (flatbed scanners) allow you to place your document on them, face down, and leave it there until they're finished with it. They are not cheap, but, if you do a lot of work with graphics applications, you may already have one. Most fax software packages don't provide direct support for scanners. However, both the *PC Magazine* Editors' Choice award winners (Eclipse FAX and FaxMaster) and two others (FAXability Plus, from Intel, and Superfax for Windows, from Pacific Image Communications, Inc.) do. If an existing document is all text, it can be processed with scanner software that has optical character-recognition (OCR) capabilities. Several packages have this capability, which can be used for scanned original documents or for incoming faxes.

BEST BET

> Unless you're both wealthy and patient, don't buy a scanner just in order to use your computer to fax existing documents—even if you will also be faxing documents you create on your computer. If you don't already have a low-cost fax machine, get one. The only scanners that are cheaper than the low-cost fax machines are of the hand-held variety, and they are awkward to use. The easy-to-use flat-bed scanners are quite expensive.

Fax software will, typically, allow you to work within whatever application you are using (word processor, drawing program, and so on) to produce the fax document. When you are ready to send the fax, you treat the fax software as if it were a printer. When you choose the print function, the software will ask you to provide cover sheet information and the number to which the fax is to be sent. It will then create a rasterized image of the cover sheet and the document, and use your fax modem to dial the number and send the fax. If the remote fax indicates a transmission error, the software will resend the fax, either at your request or automatically.

COMMON SENSE

> There is no magic to treating your fax software as a printer. For example, after you install a Windows-based fax software product, you will find a new entry in the Windows list of printer drivers. When you are ready to fax a document from one of your Windows-based applications, you choose the Windows Print Setup option (from the File Menu or the Print dialog box) and select the special fax printer driver. In reality, it's not a print driver; it's the fax program itself. But why should you care? It's made life simple for you, and that's one of the objectives of good software. Maybe it is magic, after all.
>
> For DOS, fax software uses a "capture" function, and provides a means for you to turn it on and off. When the "capture" function is on, any output to LPT1 (your computer's printer port) will be rerouted to the fax program for transmission. When it is off, your printer is, as you would expect, the destination of your printed output.

Most of the fax software that is not included in modem packages will allow you to assemble a single fax from a variety of document sources. However, Windows-based products make fax assembly much easier than do the DOS-based ones.

Receiving

Receiving faxes by computer may or may not be a good idea. If you don't want to be limited to receiving faxes by appointment, you'll have to leave your computer on all the time with the fax software's receive program loaded. There are two approaches to this. The first, which can be used whether or not you are a Windows user, is a terminate-and-stay-resident (TSR) operation. The second, which depends on windows, is a receive program that is a member of your Windows Startup Group. (If you're a Window's user, you'll know what the Startup Group is; if you're not, you probably don't care.) The disadvantage of the Windows-only approach is that you must have Windows loaded all the time if you want to receive faxes. This is less of a problem if you run Windows in enhanced mode (with a 386- or 486-based PC) than it is in standard mode (the only mode available on a 286-based PC). (In enhanced mode, your DOS-only programs can be run in a DOS window.)

WORD POWER

Terminate-and-stay-resident, or *TSR,* refers to a program that does not execute immediately, but stays in memory, to be executed only when some event occurs which triggers its operation. As long as such a program is resident in memory, other programs can run quite normally (as long as they are not too big for the remaining unused memory). For a fax receive program, the event that triggers its operation is the modem's RING message, at the COM port. A TSR fax receive program that is to be used, even when Windows is loaded, needs to be "Windows-aware." If it has just received a fax, and Windows is loaded, it will alert you (usually optionally) by the appearance of a window, containing a message. If Windows is not loaded, the alert may be in the form of a message on the last line of the screen or in a nongraphical (text only) window.

Receiving of faxes doesn't end when the modem goes back on hook (hangs up). You need to be able to read a fax; you may want to store it; and you may want to print it. Given that a fax and your computer's display are both rasterized images, displaying a fax for viewing is not a big chore for your fax software. But, what about those exasperating people who send faxes feet-first? Simple, the fax software will invert the image for you—once you notice that it is upside down on the screen. It will also invert the copy it has stored on disk.

> **PITFALL!**
>
> Not all fax software has the invert function. If, in spite of the drawbacks to receiving faxes by computer, you decide to do it anyway, make sure you choose software that has this function. It's easier than teaching the whole world not to feed documents into their fax machines feet-first. Both of the *PC Magazine* Editors' Choice fax software products can invert the image. FaxMaster can even straighten a crooked image (for those occasions when the fax looks as if the sender threw the original into the fax machine's document feed).

Caere Corporation's FaxMaster has a unique feature: the ability to convert the rasterized text in an incoming fax to an ASCII text file. As Caere's original reason for existence was optical character recognition, they have a considerable advantage, in this area, over their competitors.

Management

There are several things that can be managed by fax software: send logs, receive logs, storage (including indexing and file-compression), searching of stored faxes, and deletion of stored faxes. Oddly, *PC Magazine* treats send and receive logs as a receiving and viewing function. However, as all of the software they reviewed includes the capability to generate and maintain both kinds of log, it had little effect on their ratings. Apart from the send and receive logging functions, most fax software, with the possible exception of Eclipse fax, is fairly weak in the fax management area.

Printing

Considering that the printing of incoming faxes is a real fax machine's normal mode of operation, you would expect fax software to handle this function reasonably well. The Windows-based programs support all the printers supported by Windows (as you would expect), and the DOS-based ones support a wide range of printers; all but two of the DOS programs reviewed by *PC Magazine* support PostScript printers.

Of the fax software tested by *PC Magazine*, only the two award winners (Eclipse FAX and FaxMaster) were rated "good." The others were rated "fair" or "poor." Both Eclipse FAX and FaxMaster can be set up to print automatically on reception and print a selected page or range of pages (of a stored fax). Eclipse FAX can even print a selected area of a fax. All the Windows-based products can print in the background, but that's really a capability of the Windows Print Manager.

Other Capabilities

There are a few additional capabilities that are common to all fax software products. One is a phone book feature, which saves you from having to look up and enter numbers whenever you send a fax. Another is the broadcast capability, which allows you to send one fax to multiple recipients. Yet another is the ability to send faxes in the background, which lets you get on with other work; not all packages do this equally well. And, of course, they can all send faxes at a specified time (after midnight, if you want to save money on your long-distance bills).

What Your Computer Needs

The rule of thumb for fax software is that the Windows products require 2 megabytes of RAM and about 2 megabytes on your hard disk. The DOS products require 512 kilobytes of RAM and from 1 to 2 megabytes on your hard disk. For DOS, you can expect the TSR program for receiving faxes and for capturing outgoing faxes (from your printer port) to take about 70 kilobytes of RAM, but, for DOS 5.0 and later, you should be able to load it into high memory.

Fax Software Evolution

We've only mentioned a few fax software packages by name. In December, 1992, *PC Magazine* tested a total of thirteen. The number available is somewhat more than that. All software is evolving, including fax software. You can expect improvements in the better packages and some catching up by the others.

Benefits of Sending Faxes by Computer

As long as you are the author of your own fax documents—and don't, therefore, have to buy a scanner to convert existing (hard copy) documents—faxing by computer has several benefits, which are:

- Operation is hands-off. You don't have to feed sheets of paper into a machine.

- Paper consumption is reduced (unless you decide to make printed copies of all outgoing faxes).

- The fax software will maintain a log of all faxes sent.

- Cover sheets can be produced from a "phone book," which is a feature of all fax software.

Drawbacks to Sending Faxes by Computer

As we've discussed, sending faxes by computer is not an unalloyed blessing. There are some drawbacks, namely:

- Existing documents must be sent separately (using a fax machine) or must be scanned (expensive if you have no other requirement for a scanner).

- Illustrations must be produced with a PC graphics application (time consuming and, if you have no other requirement for it, expensive) or sketched and scanned.

- There is no equivalent to the hastily handwritten fax note (which can often be created and transmitted in one or two minutes with a fax machine).

Benefits of Receiving Faxes by Computer

There are only three clear benefits. They are:

- No need to buy a fax machine.

- The fax software will maintain a log of incoming faxes automatically.

- If you print your faxes, it will be on plain paper, rather than the curly stuff used in all but the most expensive fax machines.

Drawbacks to Receiving Faxes by Computer

The drawbacks may be major ones—even show stoppers.

- Unless you receive all faxes by prior arrangement, your computer must be switched on all the time (or, at least, during normal business hours in all time zones from which you expect to receive faxes).

- Not only must your computer be switched on all the time, but the fax software must be loaded all the time. In the case of DOS-based fax applications, this may conflict with the need to run applications which require a lot of memory.

- Incoming faxes must either be stored on disk (very space consuming, even with data compression) or printed (time consuming).

Alternatives to Fax Modems

If you don't want the bother of running your computer 24 hours a day to receive faxes, and the only faxes you want to send consist only of text and are composed by you, you might want to consider using E-mail for the purpose. All of the major online services, except PRODIGY, will allow you to send E-mail to a fax number. Some will even let you register a letterhead with them, which they will put on the cover page of every fax they send on your behalf. If you decide to subscribe to an online service, you can easily check out their fax rates and do your own cost analysis.

Evaluating Fax Software

Unless, in spite of all the drawbacks, you're absolutely determined to receive faxes on your computer, you should probably evaluate fax software solely on the basis of its fax sending, management, and printing capabilities.

If you feel you absolutely have to receive faxes by computer, you need to consider how intrusive the software will be, particularly regarding its memory requirements.

For sending faxes, consider whether you need to assemble them from a variety of sources. If you're content with straightforward text messages, you probably don't need an advanced package. On the other hand, if you're doing many creative things on your computer, and if you need to fax your output to someone on a regular basis, evaluate the fax software on the basis of how easy it makes the process of assembling faxes from multiple sources.

And don't forget the alternative for text-only messages—an on-line service such as CompuServe, which makes sending faxes as easy as sending E-mail. Even though CompuServe charges for faxes, there is never a long-distance phone charge (which you may very well incur if you do your own faxing).

Ready for the Next Step?

By now, you should be in a position to decide what you really want to use your modem for and whether or not that includes faxing. If so, you're ready for Chapter 6—unless you lead a completely unstructured life (in which case you probably wouldn't be reading this book).

Just in case you've forgotten, this chapter told you the following:

- Some modems can be used, with appropriate software, to allow you to send and receive faxes with your computer.

- If faxing is your only justification for acquiring both your computer and your modem, you may be buying yourself a very high-priced fax machine.

- Fax modems have more capabilities than other modems, but there is nothing special about their appearance.

- The standards governing the way fax machines talk to one another (and to fax modems) are Group 1, Group 2, Group 3, and Group 4. Group 1 is long dead. Group 3 fax machines can talk to other Group 3 fax machines and to Group 2 fax machines. Group 4 fax machines are just beginning to emerge.

- V.17, which defines fax operation at 14,400 bps, is good only for computer-to-computer faxing.

- The standards governing the way computers talk to fax modems are Class 1, Class 2, CAS, and SendFax. Class 2 is new and relieves the fax software of some of the burden. CAS and SendFax are only de facto standards, but CAS has a fairly wide following.

- You need to choose fax software that takes advantage of your fax modem's capabilities.

- In selecting from a range of fax software products, you need to consider four major areas—sending capability, receiving capability, fax management, and printing capability—plus some desirable extras, like optical character recognition capabilities.

- If you only need to send text by fax, you need to consider alternatives, such as faxing via E-mail.

CHAPTER 6

Planning Ahead

Don't let this chapter scare you. If you don't plan ahead, all will not be lost. You *can* come home from the store with your modem and your software, install both, unplug your phone (or the extension phone in your home office) from the wall socket, plug it into your modem, connect the modem to the wall socket with its modular phone cord, and Presto! you're ready to go!

If you intend to make your new communications setup permanent, however—and you want to live comfortably with it—you might want to do some additional planning before you go out and buy the modem. This chapter briefs you on a few additional considerations to think about in advance of buying the modem.

Where Will You Put It?

If you're planning to buy an internal modem, the answer to this question's pretty obvious. You put it in one of the available expansion slots in your computer, and you don't even have to consider the decor.

However, you do have to think carefully about "the lay of the land" when planning for an external modem. If you use the cable supplied with your modem, you have to accept that your modem will be within about five feet of your computer. If your modem doesn't come with a cable (or if you're prepared to buy your own cable anyway), you can extend the distance to as much as 25 feet. But why? There's not much point in having your modem far away from your computer.

If your computer's on your desktop, consider placing your modem beside it—but well away from where you usually put your coffee cup, teacup, beer, or whatever.

If you stand your computer on its side by your desk or table (or if it's one of those "tower" models), you might consider setting the modem on top of it.

You can even put your modem on the floor (under the desk or table, and away from your feet); if you're handy, you can mount it on the wall, next to the telephone jack.

COMMON SENSE

> Basically, you can put your external modem wherever you darned well please, just as long as it won't be stepped on, spilled on, or chewed by the dog (or the baby). If you heat your house with radiators, though, please don't put the modem on a radiator.

Overview: Adding a Telephone Line or Jack

Adding a new phone jack in your house is generally a matter of making a few simple connections. You'll need to buy the jack you want first, determine where you want to put it in the house, then connect it to the phone terminals coming into the house. Your phone company provides terminals for two lines. If you connect the new jack to the first line (which is already in use), you're simply adding another jack for that phone line. If you connect the new jack to the first line (which is already in use), you're simply adding another jack for that phone line. If, however, you connect the jack to the terminals for the second line (which is generally not in use), you're wiring the jack for a completely new line, and you'll need to call the phone company to have a number assigned to that line.

The next sections of this chapter give you more information to help you decide whether you need a new jack on an existing line or a new phone line. You'll also learn how to do the wiring for the new jack yourself.

Should You Have One Telephone Line or Two?

If you're the parent of teenagers, you may already have two lines, in which case it may be a question of whether you get a third. If you plan to use your modem more than just occasionally, you must remember that a shared line means you will be unable to make or receive normal phone calls whenever the modem is in use.

You may be on a tight budget, and unable to manage the extra monthly expense of a second line. Possibly you live alone and couldn't care less about getting and making phone calls when you're immersed in the world of your computer. But if tying up the phone will be an inconvenience to you or others in your household, and you can afford the installation (and monthly costs) of a second line, then go for it!

> **PITFALL!**
>
> When you are using your modem on a shared line, you should warn other family members not to try to make calls on an extension phone. If they do, they will almost certainly terminate your connection.

You may already have a second line to support a fax machine. In such a case, should you get a third line? Read on.

Can Your Fax Machine and Your Modem Coexist on a Line?

You *can* have a fax machine and modem attached to the same line, as long as you observe some simple rules.

If you set up your modem to receive calls (that is, put it in auto-answer mode), you must make sure your fax machine doesn't answer those calls. To do this, you can either switch the fax machine off, or you can make sure your modem is set to answer more promptly (after fewer rings) than the fax machine. You make this setting anyway when you select auto-answer mode (see Appendix B). For your fax machine, consult its manual to learn how to set the number of rings.

If you don't use software that dials calls automatically at intervals, you are in control. If your fax machine is currently sending or receiving, you can choose not to attempt a connection to CompuServe, PRODIGY, or whatever service takes your fancy.

If, on the other hand, you do use software that dials automatically at intervals (in conjunction with MCI Mail, for example), you may have minor problems. If an incoming fax is in progress when the software attempts to dial a call, there is a good chance that the modem will recognize that there is no dial tone, at which point it will send a NO DIALTONE message to the software. Typically, the software will try again later. If the modem dials "blind" (without waiting for a dial tone), or if it interprets the fax transmission signal as a dial tone (which can sometimes happen), it may affect the portion of the fax being received during the dialing process. If the fax is all 11- or 12-point text, its legibility shouldn't be affected. Smaller type or a detailed illustration could be adversely affected. Obviously, you can avoid this situation by not using the interval dialing option of such software. If you need to send a fax, you can have your software suspend interval dialing until you've sent it. (Normally such software will have functions that allow you to suspend and resume interval dialing.)

PITFALL!

> If your modem and your fax machine are on the same line and both in standby mode (awaiting incoming calls), both devices will try to answer an incoming call and the result will be unpredictable (but fear not, it won't be hazardous to your health).

If you set up your software to dial calls automatically at intervals (as you might, for example, if you subscribe to MCI Mail), having the fax in use on the same line at the time you attempt a call is not a problem. The modem will recognize that the line is in use, and will not allow dialing to take place. (In any case, present-day modems won't dial unless a dial tone is present.)

How Will You Handle Call Waiting on a Shared Voice/Modem Line?

So you're stuck with a single line, but you have Call Waiting. How will an incoming call affect your communications session? In a word, badly. It's more than likely you'll lose the connection.

What can you do about it? Something very simple. When you set up your list of phone numbers for online services or bulletin boards in your software (so that the software can dial the numbers automatically for you), precede each number with the sequence ***70** (or, if you don't have Touch Tone service, **1170**). This disables Call Waiting. Some software products allow you to enter this sequence separately, so you won't have to do it every time you add a number to a list.

Two Lines—Personal/Business Split, or Voice/Modem Split?

This book doesn't dispense tax advice; if you run a business from your home, no doubt you will have researched what is and is not deductible (with or without the help of a tax accountant). Even so, you can simplify your record-keeping by making an orderly division of use between two phone lines.

Use the first one for personal calls, plus incoming business calls. Use the second line for modem, fax, and outgoing business calls. You'll need a two-line phone in your office. See Chapter 8 (Figure 8.8) for a sample connection diagram.

If you don't operate a home-based business, it's a good idea to dedicate the first line to voice calls, and the second to the modem (and fax machine, if you have one). You don't, of course, need a two-line phone to do this.

COMMON SENSE

> If you don't want to buy a two-line phone for your home business office, you can use your fax machine's built-in phone for outgoing business phone calls.

What Does a Second Line Cost?

Including the connection fee and the installation of an extra jack (wall socket), your up-front cost will be about $100. If you're handy, you can install the extra jack (or jacks) yourself, and save about half this amount.

COMMON SENSE

> Don't involve the phone company until the internal wiring and jack installation is complete. That will allow their installer to run a test. Your contribution to the test will be to plug a phone into the new jack. (Of course, if you want, you can use the phone company as your inside wiring contractor.)

You probably won't want call forwarding and three-way calling on your second line—and you should certainly avoid call waiting—so your monthly cost should be about $14.

If you already have two lines, and want to install a third, be prepared to spend somewhat more. There is usually only a single cable from the phone company's distribution box on your street to the box on the side of your house. It contains two pairs of wires—a red and green pair for your first line, and a black and yellow pair for your second (or potential second) line.

Installing the second line usually amounts to making connections at both ends, and (possibly) replacing the box on your wall with one of the new SNI (Standard Network Interface) boxes (sometimes called a Telephone Network Interface). Installing a third line, in such a case, means laying (if you have underground services) or stringing a new cable.

WORD POWER

The *Standard Network Interface* (*SNI*) is a relatively new device (since about 1987). It is mounted on the outside of a building (such as a house), and connects the telephone company's wires to the wires that lead to the phones in the building. The typical model for a house accommodates two lines, linking their internal connections to the outside line. Each line uses a typical telephone plug and jack (RJ-11). You can unplug either line, either to make inside wiring changes, or to plug in any telephone set for testing purposes. For example, if you have a bad connection or no connection, but the telephone works when you plug it into the SNI, you'll have reason to suspect an internal wiring problem. You can open your side of the SNI by simply removing a clearly marked screw (and the padlock, if you have installed one). For obvious reasons, access to the phone company's side requires special tools.

Your phone company provides an instruction card with each SNI.

If you rent an apartment or office space, you'll have to negotiate with the landlord (who is responsible for the quality of the wiring within the building, but not for the cost of any jack beyond the first one). If you own a condominium, you may have to deal with the management committee. There may or may not be extra lines available. If you want to know what everyone's rights and responsibilities are with respect to your telephone service, call your local telephone company's business office.

Should You Plug the Phone into the Modem or the Wall?

Your modem (whether internal or external) will have two modular sockets, one marked PHONE and one marked LINE. Say you feel you can make do with a single existing wall jack. Just connect your phone into the modem's PHONE socket, and use the modular cable provided with the modem to connect the modem's LINE socket to the wall jack. When the modem isn't in use, it keeps the phone connected to the phone line. When the modem is in use, your phone should not be in use. If you use software such as Windows Cardfile to dial voice calls, the software will tell you when to pick up the phone.

If you have an additional wall jack, or if you buy a single line splitter (from any store that sells phones or phone accessories), you don't have to plug the phone into the modem. Just connect the phone to one of the splitter sockets, and the modem to the other.

How the Outlets Line Up

If you enjoy household projects, including wiring and cutting holes in your walls, the illustrations in this section will help you make the modifications needed to connect your modem, phone, and (if you have one) fax machine the way you want them.

If you're all thumbs, or otherwise disinclined to mess with things you don't understand (or don't care to understand), use the illustrations to decide what you want a handy friend, contractor, or the phone company to do for you.

If you want to live with what you have, go directly to the end of this chapter (to Test Your Knowledge).

The key to the phone wiring is to remember the following things about the red, green, black, and yellow wiring convention:

- At the phone company's SNI box, you are provided with *line 1 terminals* (identified by red and green markers) and *line 2 terminals* (black and yellow markers). If there are no markers, look at the color of the insulation on the wires that are connected to them.

- Your wall jack has four connectors. (If you look inside the socket, you'll see four tiny copper rods.) On the back of the jack assembly are four screw terminals—an outer pair and an inner pair. The outer pair is connected to the two inner copper connectors by red and green wires. The inner pair is connected to the two outer copper connectors by black and yellow wires. If you have difficulty visualizing this, just look at any of the illustrations later in this chapter.

- Each of your existing wall jacks is connected by way of a pair of wires (probably with multicolored insulation) to either the line 1 (red/green) or line 2 (black/yellow) terminals of the SNI.

- If you live in a house and it's less than thirty years old, chances are that you have a large bunch of multicolored wire pairs, leading from the SNI box to goodness knows where. Some of them are already used for the connections to existing phone outlets. Others are there for future connections. The color combination of each wire pair is unique, allowing you to find both ends in order to connect a new phone jack.

Single Line, Single Outlet

If you want to add an outlet for your modem to an existing single line, you'll need to buy a new jack from your local phone, electrical, or building-supply store.

Figure 6.1 shows a back view of a fairly commonplace outlet jack. Not all of them are circular; some are rectangular. You can connect this jack in one of two ways:

- Track down an available multicolored wire pair to make the connection directly to the SNI.

- Using a length of internal phone cable from the same store where you bought the jack, connect the indicated terminals to the corresponding ones of a nearby existing jack. (This approach is called *daisy-chaining.*) If you currently have only one line, it will be using the red/green pair. The black/yellow pair is reserved for second lines (second, fourth, sixth, etc.) only.

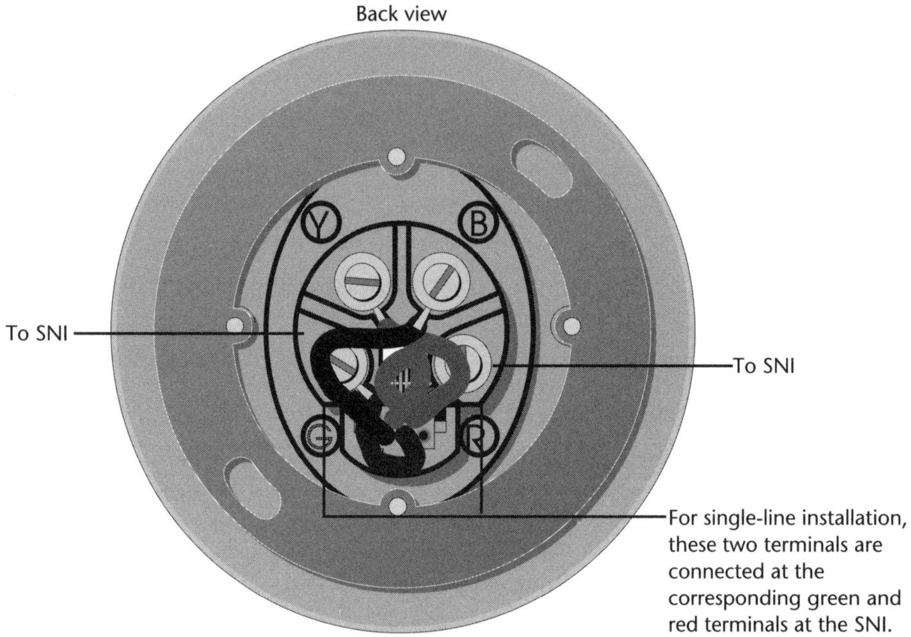

Figure 6.1
Single line, single outlet.

Two Lines, Single Outlet

If you decide to add a second line, but don't want to cut a hole in the wall for a second outlet, you can upgrade an existing outlet.

If your present jack doesn't have the terminals for the second line (some older ones have only a single pair), you'll have to buy a new one that has the appropriate terminals.

As before, the green and red terminals are connected to the corresponding pair at the SNI. In addition, the yellow and black terminals are connected to the SNI's yellow and black terminals. You can either track down one of the multicolored wire pairs for this connection, or you can buy a length of internal phone cable. Figure 6.2 shows the connections.

You will need to plug a two-line splitter into this jack. Its left socket (line 1) will accommodate your phone, and the right socket (line 2) will accommodate your modem.

PITFALL!

When buying splitters, be sure you get the right kind. When I refer to a *two-line splitter*, I mean one that plugs into a two-line jack; it provides a line-1 connection via its left socket, and a line-2 connection via its right socket. It might be sold as a l*ine 1/line 2 splitter*. A *one-line splitter*, in effect, provides you with two copies of the original socket. If the original socket supports one line, you can now plug two devices into that same line. If the original socket supports two lines, you now have two copies of the two-line socket—useful if you want to plug in a pair of two-line phones, but not much good for our purposes.

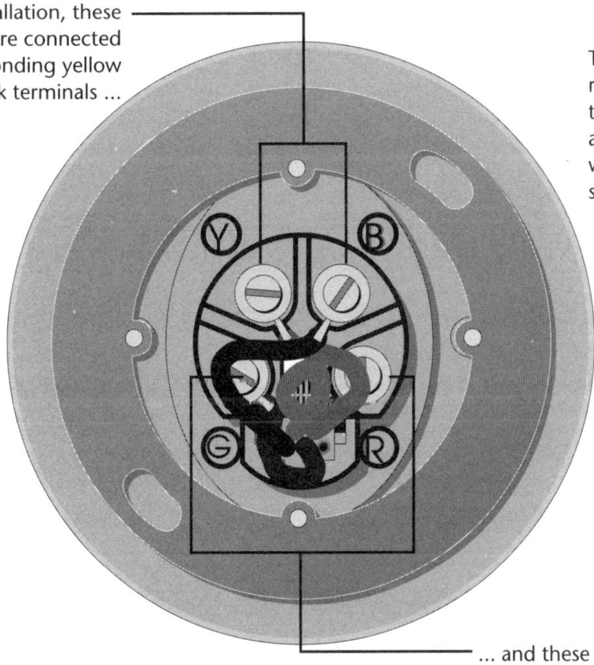

Figure 6.2
Wiring a phone jack for two phone lines.

Single Line, Two Outlets

You may not want a second line, but you might want to plug both your phone and your modem into the wall without using a splitter. In this case you would replace an existing single-jack outlet with a double one. As Figure 6.3 shows, you would simply use two short lengths of wire to connect the top green and red terminals to the bottom green and red terminals. The bottom pair would, in turn, be connected to the SNI.

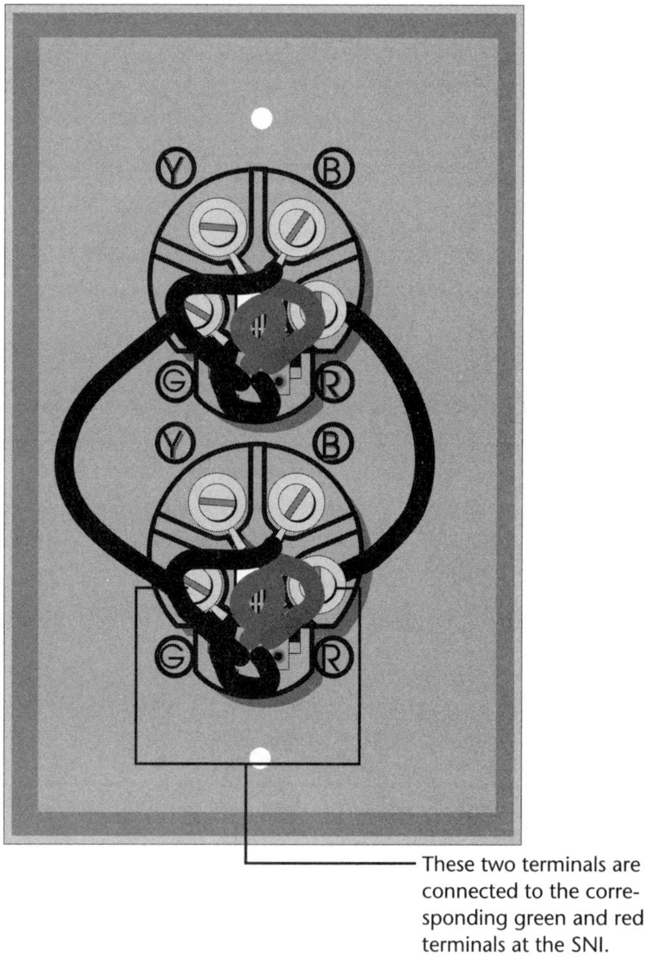

Figure 6.3
You can wire a single line to have two outlets so you can plug in the phone and the modem.

Two Lines, Two Outlets

If you already have a rectangular wall plate for a single-line outlet, or are willing to modify the hole used by an existing circular wall plate, you can provide separate jacks—on a single wall plate—for each of two lines. Because you will be plugging in single-line devices (that is, a single-line phone and modem), you need to use the green and red terminals for both line 1 and line 2. As Figure 6.4 shows, the upper green and red terminals (line 1) are connected to their counterparts at the SNI. The lower ones (line 2) are connected, not to their counterparts, but to the yellow and black terminals at the SNI.

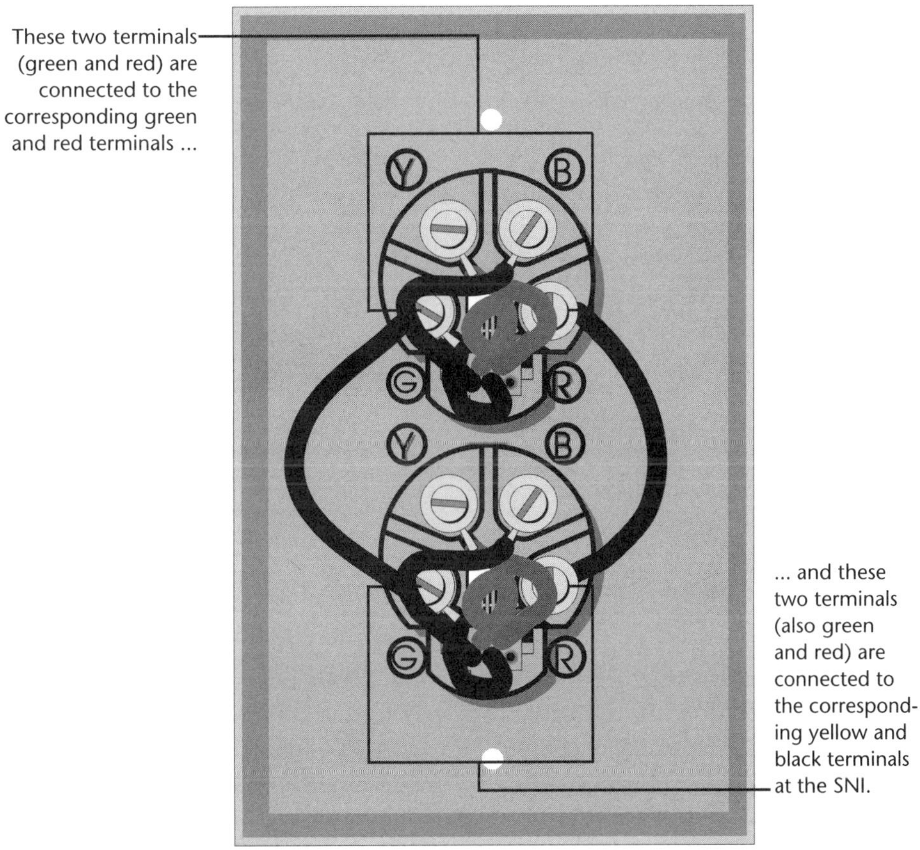

Figure 6.4
Connecting two phone lines from the SNI to two outlets on a single jack.

Creating a Home Business Setup

Figure 6.5 shows the ultimate setup, with a socket for your two-line phone, plus a pair of sockets on line 2 to accommodate your modem and your fax machine. Here, the line-2 terminals on the two-line jack are connected to the single-line terminals of each of the jacks on the rectangular wall plate. As in Figure 6.2, the two-line jack's terminals are connected to their counterparts at the SNI.

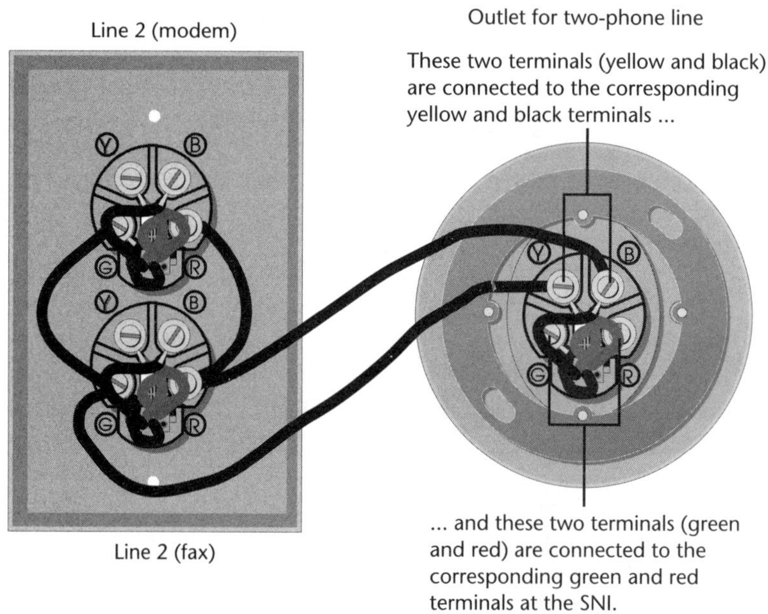

Figure 6.5
The "professional" home business setup—an outlet for a two-line phone, plus a modem outlet and a fax outlet, both on line 2.

If you're adding everything shown in Figure 6.5 to what you already have, you don't have to go all the way to the SNI for your connections. Instead, you can daisy-chain off a nearby two-line jack—that is, you can connect to the green/red and yellow/black terminal pairs of the existing jack, instead of those of the SNI. Operation of the existing jack will not be affected (as long as you don't disconnect it from the SNI inadvertently—in which case, nothing will work).

PITFALL!

Even though you've planned carefully and may have installed the extra outlets as described in this chapter, try not to install and test every device at once. Add and test your modem first. Deal with fax machines and extra phones separately. More on this in Chapter 8.

Try It! Test Your Knowledge

1. Name two disadvantages of using a modem with a single phone line.
2. If you're stuck with a single line, which of the two disadvantages can you overcome, and how?
3. Can your modem and fax machine share a line?
4. Can you do your own telephone wiring modifications?

Answers

1. Usage conflicts with teenagers or others, and interference from call-waiting signals.
2. Call waiting, by using the *70 or 1170 prefix for all modem calls. (There is no known remedy for teenagers—other than the passage of time.)
3. Yes. If your modem is turned off, or if your software has not set it to answer calls (see Appendix B), then your fax will answer. If your modem is in auto-answer mode, it will answer incoming calls, provided it is set to answer after fewer rings than the fax machine (or the fax machine is turned off). For outgoing calls, you can use either the fax or the modem, even when the other device is powered on. Just try not to be absent-minded—in other words, don't try to make a call with one while the other is in use.

4. Trick question. Only you know if you can. But you certainly may! As long as you confine your activities to your side of the phone company's Standard Network Interface (SNI) and you own your own house, there's no problem. As there are no hazardous voltages, electrical codes aren't an issue, either.

Ready for the Next Step?

Now it's time to either visit the Olde Modemme Shoppe or the Enormous Electronic Emporium, or to call Morty's Magnificent and Magical Mail-Order Company. Before you do, why don't you review your most recently acquired wisdom?

- You should put your modem where it won't be damaged.
- A single line can be a pain if your phone is already used a lot.
- Unless you're on a tight budget, two lines are usually better than one.
- There are several ways you can set up your phone jacks to accommodate your phone, your modem, and (if you have one) your fax machine.
- Two lines with a personal/business split can make your tax accounting easier.
- You can modify your own internal phone wiring if you choose to.
- A Standard Network Interface (SNI) is the point at which your wiring connects to the phone company's lines.
- A splitter may be used to duplicate the function of an existing outlet (single-line splitter), or to separate the lines of an existing two-line outlet (two-line or line 1/line 2 splitter).

CHAPTER 7

How and Where to Buy a Modem

Buying a modem is like buying any other electronic device. You have a wide choice of places to buy it, and you need to satisfy yourself that you're getting value for your money—the features you want, and some assurance that the seller will stand behind the product.

How Much Should You Pay?

If you're used to driving a hard bargain, you'll pay as little as possible. Still, it helps to know what is reasonable. In Chapter 3, you learned what the price range might be for modems of various speeds. Let's ignore the slow ones, and revisit the three that might interest you (see Table 7.1).

Table 7.1 Internal and external modem price ranges, by speed.

Maximum Speed bps	*Price Range Internal*	*External*
2400	$60 to $130	$80 to $150
9600	$170 to $370	$200 to $400
14400	$200 to $460	$220 to $500

Will you find modems that are priced outside these ranges? Of course. At the high end, you'll find modems aimed primarily at the business market. At the low end, you'll find specials, including close-out specials on the 2400-bps models. Sometimes you'll find that the internal and external versions of the same modem have the same price. That's not too common, though, as external modems have more components (enclosure, cable, power cord, power supply, indicator lights).

Bear in mind, also, that the computer accessory market is both competitive and volatile. At some point, prices must level off. Modem prices haven't yet reached that point.

What Should You Get for Your Money?

The first question here is, "What money?" Unless you're extremely well-heeled, decide what your requirements are—and what you're prepared to spend to meet them. Try to find three or four products whose cost is no more than your target price (or doesn't exceed it by more than a trivial amount).

BEST BET

Use the checklist in Appendix D to compare the modem models you're considering, or check off the features you want. Photocopy it or tear it out of the book, and take it with you.

If you are unable to find three or four modems close to (or under) your target price, you may have set a price that's too low. You can keep looking, or you can bite the bullet and move the target up a little.

Now let's see what you should get in the package, in addition to the modem. When you go shopping for your modem, check the box (or ask the salesperson) to ensure you get all the items I describe next.

Installation and User Manual

The *installation* and *user manuals* are crucial, and you shouldn't throw them away. Not only will you need them to set up the modem, but you may need them again if you ever need to remove then reinstall the modem. Better still, some modem manufacturers include a *quick installation guide* for those impatient souls who suddenly develop functional illiteracy when they see a full-size installation manual.

> **Potts' Law:** One's willingness to read manuals is inversely proportional to one's alleged intelligence. (Translation: Some people think manuals are only for dummies like us.) Keep—and read—the manuals that come with your modem.

PITFALL!

"Free" Software?

There's no such thing as a free lunch. (You just pay twice as much for the drinks.) If there's software in the package, it cost somebody something to put it there—and, ultimately, someone wants you to pay for it.

Most included software (except diagnostic software, which the manufacturer may or may not provide) is of the "let's-get-acquainted" demo variety. Typically, it lacks certain functions that, as the vendor is quick to point out, can be obtained by means of a simple (and paid) upgrade to the "professional," "commercial," or "extended" version. This doesn't mean it's useless. Most software included with modems provides some useful features, and can help you decide if you need to make an additional software purchase.

Most often, the modem package includes generic communications products (like QMODEM) and, with modems that have fax capabilities, fax software such as DosFax LITE and WinFax LITE (Delrina Technology's entry-level DOS and Windows products, respectively). Interestingly, the final step of the installation process for these products is to call Delrina's 800 number to fax your registration information to them, and (they're not dumb) to put you on their mailing list for upgrade brochures.

You will have to decide for yourself whether an upgrade is going to be worth the extra money.

Warranty—How Long?

Five-year warranties are common, and some manufacturers are offering seven-year warranties. Do you really need a seven-year warranty on something you may be replacing in a couple of years? At the rate at which advances are being made in data communications, tomorrow's modems (or tomorrow's modem replacements, whatever they may be) are sure to be faster still—possibly many times as fast—and you'll want to upgrade to keep pace (literally).

When today's electronic devices fail, it's usually because of manufacturing defects or physical damage. It's improbable that a manufacturing defect will take more than a year or so to show up. (It happens, but very rarely.) If you cause the physical damage yourself, it's unlikely that a warranty will help.

So, if the price is right, and the modem has a seven-year warranty anyway, you can welcome the extra "peace of mind" protection. If, just to get such a warranty, you have to buy a more expensive modem than you would really like, it isn't worth it. If the salesperson tries to sell you an extended warranty, politely and firmly decline. It's better, anyway, to use a credit card that doubles the length of the manufacturer's warranty (subject to a one-year maximum increase). If you have such a credit card, the issuer should already have explained the benefit to you.

Will You Get an Online Service Software Package?

Sometimes you get a modem with software included. Sometimes, the package offers software with a modem thrown in.

PRODIGY is one online service provider that has, from time to time, offered a membership kit that included a modem. In 1989, one of their better deals included a Hayes Personal Modem 2400. Was it good quality? Yes. Was it good value? At the time, yes; it provided the means to connect to the service, and also worked perfectly well for CompuServe, America Online, and numerous other services. Would it be good value today? At not much under $200, which was how the kit was priced then, no!

BEST BET

> Where an online service company offers a membership kit that includes a modem, evaluate the modem separately. Subtract the price of the basic kit to determine what they're charging for the modem.

Where Should You Buy It?

You have numerous choices of where to buy your modem. But, to acquaint yourself with the full range of possibilities, you should check your local Yellow Pages for Computer Retailers, your local newspaper's weekly Computing supplement (if it has one), the computer magazines, and the free computer newspapers like *Micro Times* and *Computer Currents*. (Many local bookstores sell computer magazines.)

Following is more information about some of the sources that offer modems. Don't limit yourself to just one. Shop around.

Buying Through Mail Order

The major mail-order houses place multipage advertisements in every issue of computer publications like *PC Magazine, PC World, PC Computing,* and so on. In general, mail order computer suppliers are trustworthy, stand behind the products they sell, and compete quite well on price. Most will give you a choice of surface shipment or (at extra cost) one- or two-day air shipment. Read the fine print in their ads regarding returns, possible restocking fees, and so on.

PITFALL!

> The two major disadvantages of mail order are that you have to pay for shipment (both ways if you return something), and that you don't get to see and handle the product before you buy it (although you could do that in a store, then order the same item from a mail-order house).

Shopping at Computer or Office Superstores

Most metropolitan areas have several computer or office megastores. These stores benefit from volume discounts from manufacturers. Usually, they'll pass the benefit on to you, in the form of lower prices. Often, they'll match the lowest retail price in the area for the same item. Typically, they won't match the lowest mail-order prices, but they often have reasonable return policies and very liberal exchange policies.

COMMON SENSE

> When you go into your local megastore, look at the customers at the return counter. In general, do they seem to be contented or agitated? Stand close enough to overhear what they and the store's customer service people are saying. If you don't like what you hear, take your business elsewhere.

Casing the Smaller Computer Stores

Indoor malls will typically have one or more franchised or chain computer stores, such as Radio Shack, Software Etc., and Egghead Software (which also sells by mail order). Even the ones that call themselves software stores may sell accessories such as modems. They usually have a small and reasonably knowledgeable (but don't count on it) staff.

Outdoor malls, including the neighborhood strip malls, also have the better-known chain and franchise stores. Often, though, they'll also have a locally owned or regional computer specialty store. Such stores are more likely to have someone knowledgeable to help you, but find it a little more difficult to compete on price.

Checking Out Computer Stores That Sell to Businesses

You'll recognize a computer store that sells to businesses when you walk in. It will, very likely, look like a cross between a showroom and a bookstore, having a small selling area with a counter (and possibly some sales desks) nearby. The range of goods may be limited, and discounts may be offered only for large-quantity purchases.

These stores probably want your business as much as the next store, however, and may be quite amenable to price negotiation. Also, there's a reasonable chance of talking to someone with some knowledge of the product you're looking for (modems, in your case, just in case you forgot).

Comparison of Sources

It's hard to recommend one type of place over another, but the summary in Table 7.1 provides you with some very general guidelines—but remember, there are always exceptions. You must decide what's important to you.

Table 7.2 Advantages and disadvantages of modem sources.

Source	*Strengths*	*Weaknesses*
Mail order	Low prices (usually) Fairly wide range	Shipping costs No personal service Limited expertise
Computer superstore	Low prices (but higher than mail order) Moderately good range	Limited personal service Variable expertise
Office superstore	Low prices	Limited personal service Very little expertise Limited range
Small store (chain or franchise)	Convenience Low prices Personal service	Limited expertise Limited range
Computer specialty store	Expertise Personal service After-sales service	Less competitive prices Limited range
Business computer store	Expertise Personal service	Limited range Pricing might be attractive only for quantity sales

Dealing with the Salesperson

Some computer salespeople can be very inexperienced, particularly if they're working in an office superstore, where computer knowledge isn't necessarily the most important job requirement. There are also many who do know what they're talking about.

If all the salesperson does is point you to the right shelves or recite a memorized list of the modems the store sells, ask to speak to the manager of the modem department. There won't, of course, be a modem department, but you should get the responsible manager anyway. If the manager's not available, ask for a specialist. If you don't get either, take your checklist (Appendix D) over to the shelves and look at the specifications and prices on the boxes anyway.

In smaller computer stores, you'll find it easier to locate the experts. High-pressure sales tactics are rare in computer stores, thank goodness. In the rare event that someone does turn on the pressure, resist. Don't be rushed.

Ask Questions

"Those who do not wish to know ask questions of those who cannot tell." (Raleigh, 1861-1922.)

You *do* wish to know about modems, so once you've found someone who seems to know the answers, *ask questions*. Take out your checklist (Appendix D, remember) and ask if the store has any modems that meet all your requirements (or if the one that's caught your fancy is all it claims to be). Ask, especially, what the store's return and exchange policies are.

> If you think you have a dumb question, ask it anyway. Actually, there are no dumb questions, only dumb answers. If you keep getting dumb answers, ask someone else.
>
> **COMMON SENSE**

Most Important Questions to Ask the Salesperson

You may be able to answer some of these questions yourself, by reading the print on the modem package. The questions that relate to store policy and practices will require the help of a salesperson.

- Is this the best modem in its price range?
- May I see the manual?
- Is a modem cable included?
- Is a 9-pin-to-25-pin adapter included? (Ask this only if you need one!)
- What is the store's policy on exchanges and refunds?
- Does the store provide out-of-warranty service? How is such service priced?
- Does the modem have 16550 UARTs? (Internal modem only—there may be a significant performance advantage when the modem is using MNP Class 5 or V.42 *bis* data compression. See *UART* in the Glossary.)

How Should You Pay?

If you know the store or know people who are satisfied with the store's policy on refunds, it doesn't matter how you pay. If you want to hedge your bets, charge it. If the product is unsatisfactory and you have trouble getting satisfaction from the store, you can appeal to the issuer of the credit card (see the back of your statement).

Also—and it doesn't hurt to say it again—some credit card issuers will extend the manufacturer's warranty by up to a year on anything you purchase (at least with their "preferred" cards). Chances are you won't need it, but if it costs no more ...

BEST BET

Old Yorkshire Saying: "If there's owt fer nowt, git it fer thee sen." (Translation: If there's anything for nothing, get it for yourself.) Take advantage of your credit card's services when buying your modem.

Ready for the Next Step?

This chapter and the preceding ones helped you decide what kind of modem to buy, how much to pay, and where to buy it. Once you get it, you'll have to install it and get it running. (The next few chapters will help you with installation and setup.) But before you read about that, let's review the main points of this chapter.

- Prices vary widely, and you should aim low.
- You should look for a modem that comes with good documentation.
- You should look for some "free" software with your modem. If it's a fax modem, it should come with entry-level fax software.
- It's nice to get a long warranty, but you shouldn't pay premium prices for it.
- If a modem is included with an online service's membership, you should evaluate the modem as a separate item.
- You now know the kinds of places that sell modems. You need to shop around before you buy.
- You should ask questions, including the ones listed.

CHAPTER 8

Installing Your Modem

Murphy's First Law states that if anything can possibly go wrong, it will—and at the worst possible time. Robert Burns, the Scottish bard, put it more succinctly with "The best laid schemes o' mice an' men gang aft a-gley."

Having come this far, you can prove both Murphy and Burns wrong. Follow this chapter, and you'll have no problems installing your modem.

Internal Modem

If you've installed other circuit boards in your computer, this will be a breeze. If you haven't, you should know that you may need some simple tools—especially one or more screwdrivers. Keep a small flat screwdriver and a small Phillips screwdriver handy. You might possibly need a screwdriver with a hexagonal head. In any case, check the documentation that came with your computer. It will specify the type and size of any required screwdrivers.

If you keep your computer on your desk or tabletop, you can probably work on it in place. The same is probably true for a "tower" model, as long as it isn't wedged between two other objects. Otherwise, you may have to work on an available table, a workbench, or even the floor.

Read the Manual!

Manuals are especially important with internal modems. This is because you have to set them up before you install them. So, before you proceed, read the manual that came with your modem, and follow the steps it gives you for setting the modem up. (Then keep the manual handy while you continue with the installation.)

What Are DIP Switches, and Do You Need to Set Them?

The most common component you adjust during setup is the *DIP switch*. Figure 8.1 shows four switches in one DIP assembly. It would normally be referred to as a *four-position* DIP switch. Some switches have a rocker action, some have a slide action, and still others have an up-and-down action. No matter how a switch works, its role in life is simple—it's either on or off.

> **WORD POWER**
>
> A *DIP switch* is a switch in a "dual inline package" assembly, which has two rows of connectors that plug into a circuit board. You use DIP switches to make setup choices, such as COM port (COM1, COM2, COM3, etc.).

Your modem may have a DIP switch with more or fewer than four positions. It may, instead, have metal jumpers used to connect adjacent posts that stick out of the printed circuit board. It may have a slide switch with several positions, similar to the OFF-LOW-HIGH switch on your phone that sets the ringer. What's important is that the installation manual will describe the switch(es), almost certainly with an illustration, and tell you what to do. Take care of this task first.

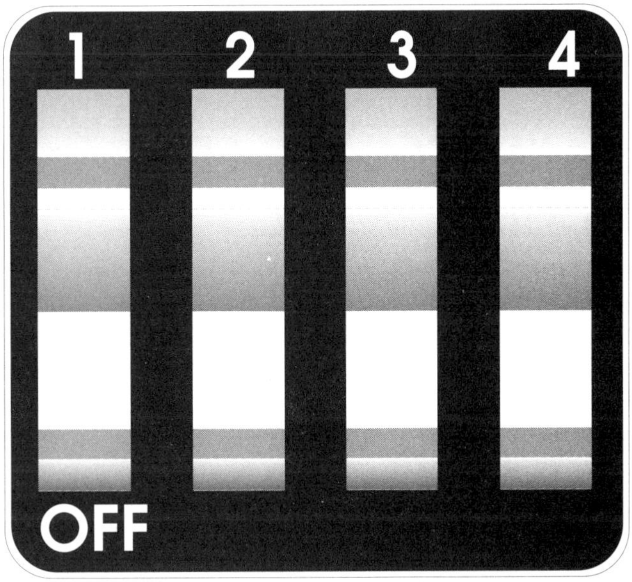

Figure 8.1
Four-position DIP switch.

The most common DIP switch settings relate to the number of the COM port you will be using. Most internal modems will give you a choice of COM1, COM2, COM3, and COM4. Some will even go as high as COM8. If your computer has COM1 and COM2 built in, you will almost certainly have to choose COM3 or higher. Don't choose COM5 or higher unless you're sure all the communications software products you'll be using support ports higher than COM4. Most don't, but such support is becoming more common.

> **PITFALL!**
>
> As you set DIP switches, read the instructions carefully. Some DIP switch settings may relate to *IRQ* (interrupt request) numbers. If so, you'll want to avoid conflicts, especially with a *bus mouse* if you have one (this is a mouse that connects directly to a circuit board within your computer, instead of a port; customarily it will use the interrupt request IRQ 5).

There may be DIP switch positions that are used for operation in *test mode*. Typically this is a manufacturing test, but it may relate to a test that a user can run. In any case, make sure you set such switches for normal operation.

Opening Your PC Without Damaging It (or Yourself)

If you've never opened your personal computer before, now's the time to learn, or call in an experienced friend.

First, disconnect the power cable, and if the monitor sits on top, move it to one side. Move the keyboard to one side too (as most covers are pulled off towards the front). This procedure will be somewhat different for a "tower" model.

Next, remove the computer's case to expose the internal parts. Most computers now have knurled screws that you can turn with finger and thumb, or large ones designed to be turned with a coin. With other computers, including older ones, you may need a screwdriver (flat, cross, or hex). Follow the computer manufacturer's directions to remove the screws and pull off the case. Remember to put all the screws in a safe location; you'll need them to secure the case when you put it back on the system unit.

Once you have removed the cover, look for an empty slot in which to install your modem. *Slots* are black plastic trenches lined with metal, into which you plug circuit cards such as modems and video adapters. Locate a slot that is the same length as the metal-covered tab at the bottom of your modem circuit card. If you need to, you can use a slot that's longer, but not a slot that's shorter.

BEST BET

> There are three sizes of slots: 8-bit (one connector strip), 16-bit (two connector strips, end to end), and 32-bit (yet another connector strip). Most modems can use an 8-bit interface, but can go in any of the three slot types. However, if an 8-bit slot is available, there's no point in using up one of your 16- or 32-bit slots, which you might want to keep available for future add-in boards.

After identifying the slot to use, remove its backplate. The *backplate* is the metal cover that keeps dust out of the computer when a slot is empty; it needs to be removed so cords (for instance, the phone cord) can be plugged into the card you're installing. To remove a backplate, remove the screw at the top, then lift the plate out. Be careful not to drop the screw inside the computer—save it to secure the modem in place. Save the backplate in case you ever want to remove the modem.

Inserting and Securing the Modem Board

Position the modem over the slot you've chosen for it, then push the modem firmly but carefully into place. If it's a tight fit, seesaw the ends of the card until it is completely seated. Secure the modem's backplate with the screw you took out to remove the blank backplate.

> **PITFALL!**
> Never force your modem (or any other board) into place. You could damage either the board or your computer.

Do You Replace the Cover Immediately?

If you're going to check the modem out right away (by installing your communications software and making your first online connection, as described in Chapter 9), you might want to leave the cover off for awhile. If that makes you nervous, by all means replace it immediately. When you do replace the cover, do so carefully, according to the manufacturer's instructions, and with the power disconnected.

Connecting the Phone Cables

In Chapter 6, we described some of the things you could do with your phone wiring, wall jacks, and so on. If you made any changes, you might want to review the relevant illustrations in Chapter 6.

If you're using an existing single-line phone jack, make your connections as shown in Figure 8.2. This illustration is somewhat simplified; many internal modems have a square opening on the backplate for access to the

DIP switches. Note that the phone cord is connected to the socket marked PHONE and the cord from the wall jack is plugged into the socket marked LINE. (Some modem manufacturers use the word TELCO, for telephone company, instead of LINE.)

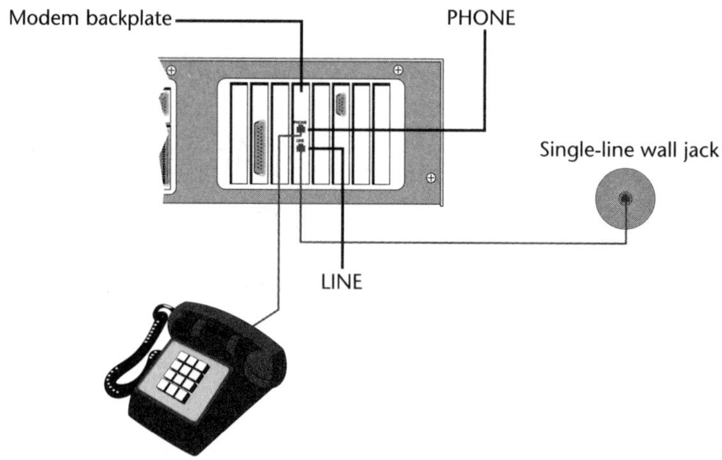

Figure 8.2
Connecting your internal modem to a single-line jack.

If you would rather not plug the phone into the modem, you can use a single-line splitter at the phone jack, as shown in Figure 8.3. Functionally, it's no different (and you still have only one line).

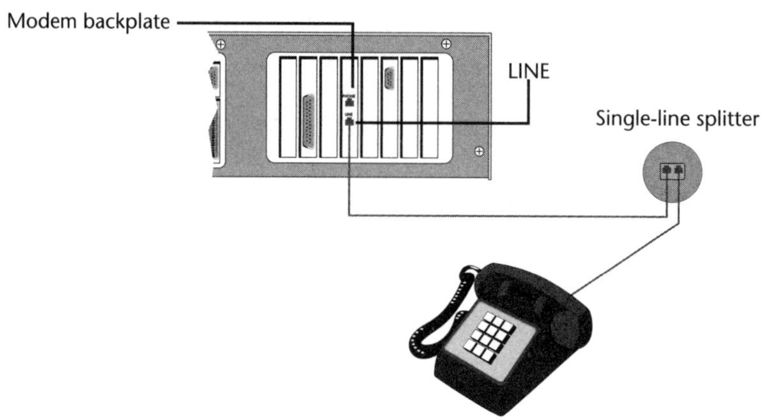

Figure 8.3
Using a splitter so that you don't have to plug your phone into your modem.

If you have a two-line, two-jack outlet (see, also Figure 6.4), your connection will be as shown in Figure 8.4. The inset shows the alternative jack arrangement (corresponding to Figure 6.2) with a single-jack, two-line outlet. This arrangement lets you use the phone and modem simultaneously.

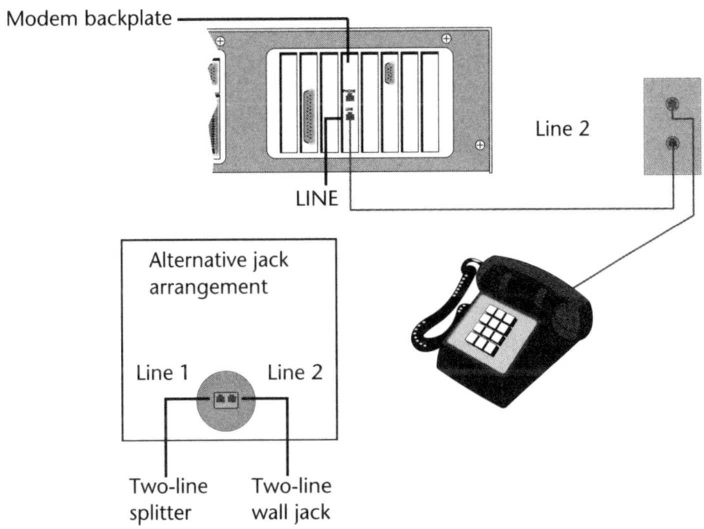

Figure 8.4
Connecting your modem and phone to separate phone lines.

External Modem

Unlike the internal modem, an external modem cannot get its power from your computer. Thus, it will have a power cord—and, probably, a combined transformer and plug (see Chapter 7). Apart from that, an external modem requires no "installation."

Read the Manual!

Manuals are less important with external modems. But read the installation manual anyway! Even external modems may have DIP switches. Only the manual will tell you what they do, and how to set them.

Connecting the Modem to Your PC's COM Port

Your external modem will need a cable—specifically, a *serial cable*, so called because it lets devices communicate with your system through the *serial ports* (another name for COM ports). There are three possibilities:

- Your modem may come with a separate adapter cable (with a 25-pin male connector at one end, and a 9-pin female connector at the other).
- Your modem may have an attached cable with a 9-pin female connector on the end.
- You may have bought your own cable (probably at the same place you bought the modem).

If your computer's COM ports (clearly marked COM1 and COM2) have 9-pin connectors where you plug in cables, you can use the cable just described (the one that's provided with the modem). If the COM ports have (or the available one has) 25-pin connectors, you'll need to use the 9-to-25-pin adapter almost certainly included with the modem. (If the adapter isn't included, you can buy it where you bought the modem, or at almost any electronic supply store.) If you bought your own cable where you bought the modem, it will (if you followed the checklist) have a 25-socket connector at the PC end.

Connect the cable to the COM port you'll be using, and to the modem. Tighten the screws on the connectors.

Connecting the Phone Cables

If you're using an existing single-line phone jack, make your connections as shown in Figure 8.5.

If you would rather not plug the phone into the modem, you can use a single-line splitter at the phone jack, as shown for the internal modem in Figure 8.3.

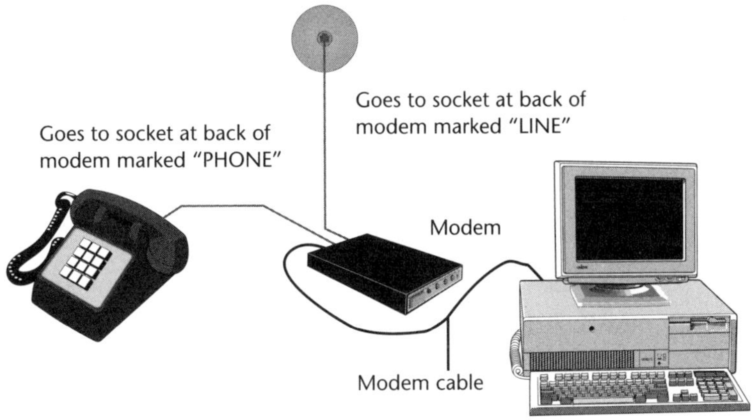

Figure 8.5
Plugging an external modem into a single-line jack.

If you have a two-line, two-jack outlet (see, also Figure 6.4), your connection will be as shown in Figure 8.6. The inset shows the alternative jack arrangement (corresponding to Figure 6.2) with a single-jack, two-line outlet.

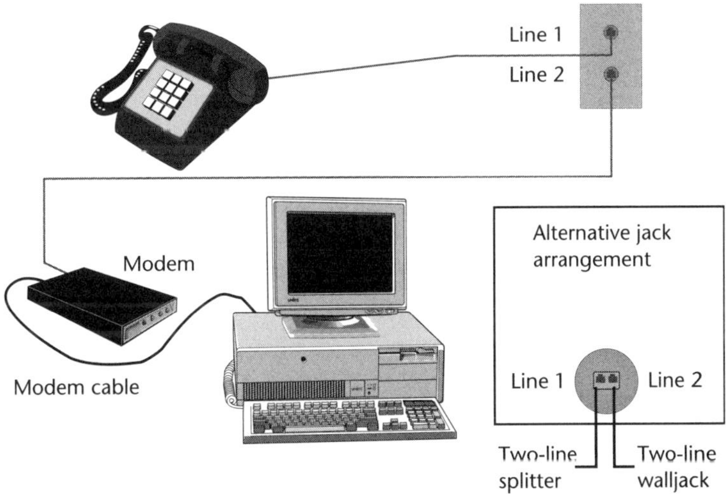

Figure 8.6
Setting up an external modem when you have two phone lines.

If you want the full business setup with fax machine, described in Chapter 6 (Figure 6.5), you can make the connections shown in Figure 8.7. You can, of course, do the same thing with an internal modem.

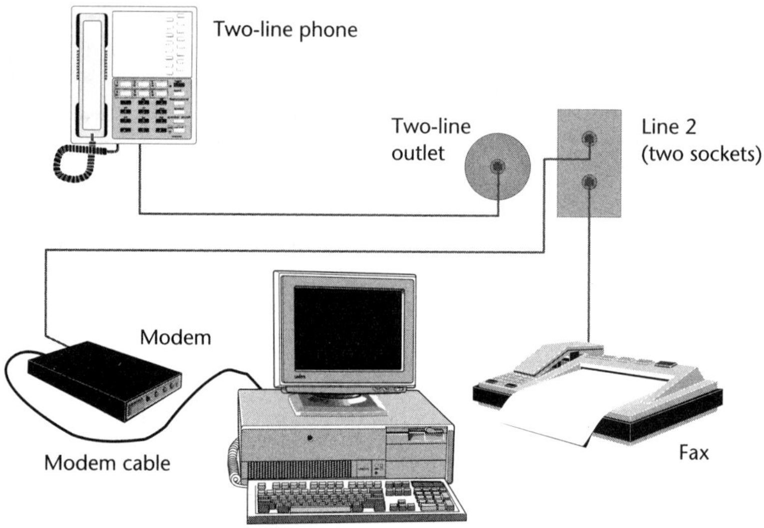

Figure 8.7
Two-line connection with fax machine.

Note that it's a good idea to wait until you have tested the modem before plugging the fax machine into the wall jack.

Finally, plug the modem's power cable (or transformer plug) into the nearest wall socket (or, if you have one, your power strip) and turn it on.

CHAPTER 9

Getting It Running

At the end of Chapter 8, you read about reconnecting power to your computer, and (for an external modem) making sure it's switched on.

This chapter uses two examples to show you, generally, how to set up the modem software. Use these examples to get a feel for what you'll be doing. For the real thing, use the instructions provided by the software vendor and the online service provider.

Installing the Communications Software

As when you installed the modem, read the installation procedure in the manual that came with the communications software (if there is one) before you install the software. This will help you get a feel for the task before you start.

If you're installing a Windows-based product, you will almost certainly follow the usual procedure, which is to place the first (or only) distribution diskette in drive A or B, choose **R**un from the Program Manager's **F**ile Menu, and enter the command **a:setup** or **b:setup** (alternatively, **a:install** or **b:install**). The product's setup program will prompt you for drive and directory information, and install the program in its own Program Group. You can then choose to leave it in the original group, or drag its icon into another group if you prefer.

If you're installing a DOS product, you will probably put the first (or only) distribution diskette in drive A or B, change to an **a:** or **b:** prompt, and enter a command such as **install**. The procedure may be different from this, so read the manual!

Configuring the Software

We couldn't possibly tell you how to set up every available software package. The two descriptions presented here, however, should give you a good idea of what you have to do. They should also give you some reassurance that it's not too difficult.

Windows Terminal Setup

If you have Windows, you also have Windows Terminal. Windows Terminal is a generic communications program. The settings you consider for Terminal will be the same as those you need to consider for other generic communications packages. The following setup steps should allow you to get started with an online service or a bulletin board. The illustrations apply to either Windows 3.0 or 3.1.

Entering the Phone Number to Dial

Start Windows Terminal by opening the **Accessories** program group and selecting the **Terminal** icon. Setting up the phone number for Terminal is easy. Just select the **S**ettings pull-down menu (see Figure 4.1), and select Phone **N**umber (or press **Alt-S** and the **Enter** key). Figure 9.1 shows the dialog box you'll see.

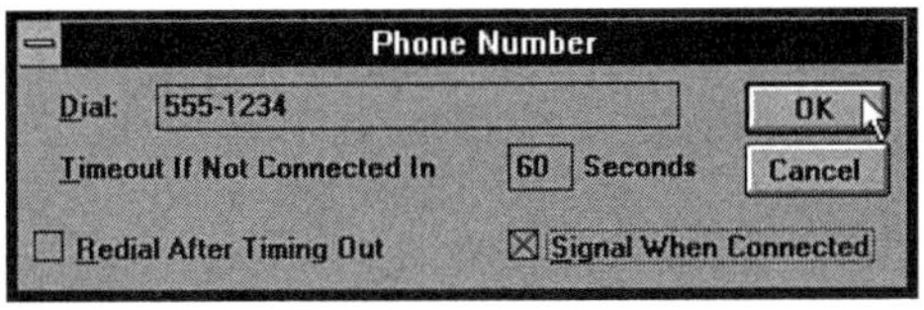

Figure 9.1
Windows Terminal Phone Number Setup.

If you will be using one of the major online services (other than PRODIGY and America Online, which require special software), you will probably enter the 800 number provided in their membership kit (or, in some cases, their advertisements). Once you are registered, they will provide you with a local number. For other services, your kit may already provide the local number.

Also in the Phone Number dialog box, the **T**imeout If Not Connected value specifies how long Terminal is to wait for an answer after dialing the number, and also how long it is to wait before redialing if you choose the **R**edial After Timing Out option. If you specify 30, for example, it will dial (when you tell it to do so), wait 30 seconds for an answer, hang up, wait another 30 seconds, then restart the process. (It allows you to cancel at any time.) If you don't specify redialing, it will give up after 30 seconds. It may seem futile to redial when there is no answer; however, as Terminal doesn't look for the modem's *BUSY* message, you may want an automatic retry in case a busy line is the reason for the absence of an answer. Choose the **S**ignal When Connected option if you want Terminal to beep when it gets a *CONNECT* message from the modem. Choose **OK** when you're finished with this dialog box.

Specifying Terminal Emulation

Terminal Emulation is the next option on the **S**ettings menu. When you select it, you will see the three choices shown in Figure 9.2 (Chapter 4 discusses these emulations). Windows Terminal offers us the DEC VT100 as the default, because of its status as a standard. Make your choice, and click on **OK** to confirm it.

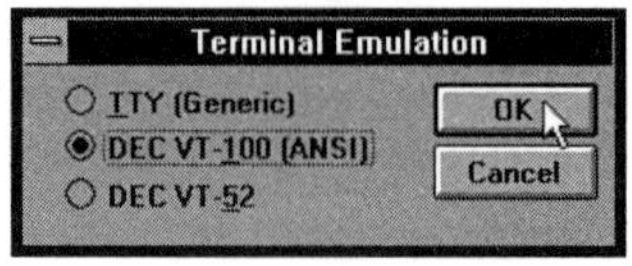

Figure 9.2
Windows Terminal Emulation Setup.

Some online services (DELPHI, for example) will ask you for your terminal type when you sign on. For DEC VT100 emulation, you would simply type **VT100**.

Next, there are important settings relating to how Terminal will communicate, which COM port it will use, and so on. To make these settings (or to determine that you want to accept Terminal's default settings), you'll need pull down the **Settings** menu again and select **Communications**.

Choosing the Right Communications Settings

At first, the Communications dialog box seems a little daunting, with eight different things you need to specify. However, you can generally accept the default setting for everything except Connector, Baud Rate and, possibly, Flow Control.

- For Connector, you must specify the COM port to which your external modem is attached, or to which you have set your internal modem. Terminal allows you to select COM1, COM2, COM3, or COM4.

- For Baud Rate, which is the term used by Microsoft and others to signify *bit rate* (also called *data rate*, and expressed in bits per second), enter your modem's maximum speed (if the speed supported by the online service for which you will be using Terminal is lower, enter that speed instead). Alternatively, if your modem supports MNP Class 5 or V.42 *bis* data compression, select **19200** and, under **Flow Control**, select **Hardware**.

- Data Bits, Stop Bits, and Parity are all related. In almost all cases, you can accept the defaults—8 data bits, 1 stop bit, and no parity; most neighborhood bulletin board systems use this combination. CompuServe uses 7 data bits, 1 stop bit, and even parity. Online services will usually provide this information in their ads or their membership kit. Just use whatever settings they specify. (Later on, you may want to look up *data bits, start-stop transmission,* and *parity* in the Glossary.)

- Don't worry, for now, about the Parity Check and Carrier Detect settings. If you're interested, the Windows User's Guide describes how Terminal uses them.

When you're finished, click on **OK** to accept your settings.

Finally, you'll need to select Modem Commands from the Settings menu. This displays the dialog box shown in Figure 9.3, which shows that you need to use the Hayes Command Language to specify some settings in Terminal.

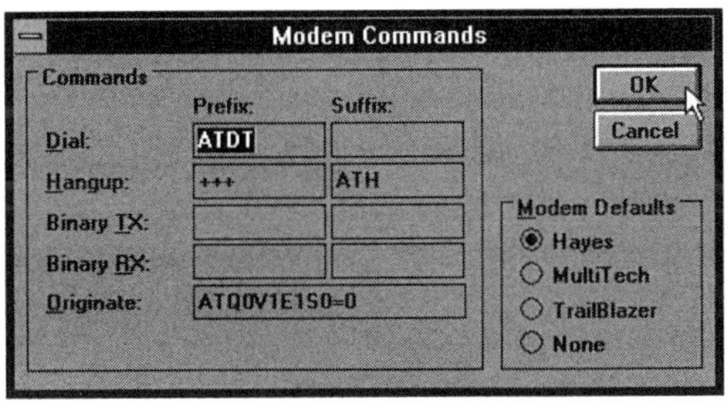

Figure 9.3
Windows Terminal Modem Command Setup.

Understanding the Hayes Commands

The commands shown in Figure 9.3 are the default commands for most Hayes-compatible modems. Let's look at each one in turn.

AT is the attention command. In effect, it says, "Hey, modem. Listen up!" It is used as part of three other commands in the dialog box to tell the modem to pay attention.

ATDT (the **Dial** prefix) is sent to the modem to prepare it for dialing. When you select **Dial** from the **Phone** menu, Windows Terminal sends the characters ATDT and the phone number (**ATDT5551234**, for example) to whichever COM port you have specified.

COMMON SENSE

If your modem happened to be sharing a line that had the Call Waiting feature, you would add ***70** to the ATDT (giving you **ATDT*70**), so as to disable call waiting whenever you use your modem. (If you want, you could add *70 to the beginning of the phone number instead. Either approach will work.)

DT indicates tone dialing. If your local exchange has only rotary (pulse) dialing support (a very rare situation these days), you can change the T to a **P** (for pulse). Including the disabling of call waiting, you would replace the default characters with **ATDP1170**. For good measure, you might want to add a comma after the **1170**, providing a 2-second pause in case there's a delay in getting a second dial tone.

If you wanted to set up dialing for international calls, you might set up a Prefix of **ATDT011** and a Suffix of **#**. The "011" is the international dialing prefix. The use of a pound sign (#) at the end of an international number indicates that the number is complete, and that the phone system shouldn't wait for any more digits.

The **H**angup Prefix **+++** will switch the modem from data mode to command mode, and the Suffix **ATH** will tell the modem to hang up. Once you have established a connection, the modem switches to data mode automatically. If you want to hang up, you have to first switch the modem into command mode. The Hangup prefix and suffix shown here will be sent to the modem when you choose **H**angup from the **P**hone menu.

WORD POWER

Data mode is the mode in which your modem transmits and receives data. *Command mode* is the mode in which it accepts and executes (obeys) commands (such as D for dial, T for tone, and H for hangup).

The string of characters shown next to **O**riginate tells the modem how it is to handle certain functions. Although this string is at the bottom of the list, sending these characters to the modem is the first thing Windows Terminal will do when you ask it to dial the phone number you entered.

Q0 (Q followed by zero) tells the modem to send responses to your computer. Modem responses are used to indicate the correct execution of commands, or that they have failed. Some communications software products require the responses in order to take action on your behalf. Windows Terminal simply displays them on the current line of your terminal screen.

V1 indicates to the modem that its responses are to be in the form of words (such as OK, ERROR, and so on). For Windows Terminal, this is the obvious choice. Communications products that do something with the responses might use V0, indicating that response numbers are to be used.

E1 says that command characters are to be echoed. You will normally set up terminal emulation to run with Local Echo off (using the **Terminal Preferences** option of the **Setup** menu). With local echo off, you see characters you type only because the remote computer sends them back to you. E1 tells the modem to do the same for modem commands as the remote computer does for data. This is useful if you are going to enter modem commands at the keyboard yourself.

Finally, **S0=0** is used to set one of the modem's S registers.

WORD POWER

An *S register*, in a Hayes-compatible modem, is a special memory location in which an 8-bit (1-byte) value is stored for a particular purpose. Some of the registers and the purposes for which they are used are described in Appendix B. Your modem's user manual will describe the ones supported by your modem.

Register **S0** indicates whether the modem is to answer incoming calls and, if so, how soon. **S0=0** sets the modem's auto-answer to off—that is, it tells the modem not to answer calls. **S0=1** would put the modem in auto-answer mode, and instruct it to answer on the first ring; **S0=3** would instruct it to answer after three rings; and so on.

Other Settings

There are other options in the Settings menu, allowing you to control certain aspects of terminal emulation (Terminal Preferences and Function Keys), to specify how you want text file transfers to be handled, and to select the protocol to be used for binary file transfers. You can probably skip these until after you've used Terminal to engage in some initial exploration of the online service for which you'll be using it. When you are ready, the Windows User's Guide will provide you with the necessary guidance for using these options.

Saving Your Settings

In Chapter 4, we told you about using the Save As option of the File menu to save your newly-modified version of Terminal under an appropriate name (**cserve.trm**, for example). If you forget to do this, Windows will ask you if you want to save your changes when you finally close the Terminal window.

Starting and Ending an Online Session

To start an online session with Terminal, either use your mouse to select the Dial option on the Phone pull-down menu, or press **Alt-P**, followed by **D** or **Enter**. Once a CONNECT message appears, proceed as instructed in the documentation provided by the online service.

When you want to disconnect, follow the online service's documented signoff procedure, then select Hangup from the Phone menu (with your mouse, or by pressing **Alt-P**, followed by **H**). Then close the Terminal Window.

Checklist of Settings

Here's a quick list of the essential settings. If you use generic software other than Windows Terminal, you will be asked to enter or verify the same kind of information. For non-essential settings, you can read the software manual at your leisure, then make whatever changes seem helpful to you.

- **Phone number:** Make sure it's the right one. If you get no answer, it's probably wrong. If a voice answers, it's definitely wrong, and you've also disturbed someone.

- **Terminal Emulation: VT100 (ANSI)** is the best bet. If you sign on to a bulletin board system that requires you to use VT52 emulation, you can select **VT52** instead. It's very unlikely that you'll need to use TTY (Generic).

Modem Commands

- **Dial Prcfix:** ATDT for tone; ATDP for pulse.

- **Dial Suffix:** Leave blank.

- **Hangup Prefix:** +++ is the default. Accept it.

- **Hangup Suffix: ATH** is the default. Accept it.

- **Binary TX and Binary RX:** These are not required for Hayes-compatible modems. Leave the Prefix and Suffix blank.

- **Originate:** The default string (ATQ0V1E1S0=0) is usually adequate. Your modem manual may have some advice about additional commands that can be added to the string.

Communications

- **Baud Rate:** Use the published rate (in bits per second) for the online service phone number you'll be using; if your modem has MNP Class 5 or V.42 *bis* data compression, select **19200** (and, under Flow Control, select **Hardware**).

- **Data Bits, Stop Bits, and Parity:** Use the values specified by the online service you'll be using (almost certainly 8 data bits, 1 stop bit and None).

- **Flow Control:** The default of Xon/Xoff is the safest choice, except as described for Baud Rate (above).

- **Connector:** It's absolutely vital that you get this one right. If you have an external modem, the default (COM1) is your most likely choice. If you have an internal modem, COM3 is the most likely choice.

Setting Up Online Service Software

As you've already learned, many online services provide you with special communications software for gaining access to the service. This section explains how you can install and set up this kind of software, using PRODIGY as an example. The Prodigy Service aims its product at a wide audience, and—to make things easy—provides you with very simple and limited setup options.

First, you need to install the PRODIGY software on your computer. The PRODIGY Start-up Kit provides you with the necessary instructions to do this and to start PRODIGY.

Figure 9.4 shows the initial screen, which gives you a choice. You may elect to use the service by entering your ID and password (both of which are provided in the PRODIGY Start-up Kit). The software will make the call for you, and use that information to sign you on; then it will present an initial screen similar to the one shown in Figure 4.3 (Chapter 4). If you have never used the service before, or need to change something, you may choose the >SET-UP option. To do so, you either move the mouse pointer to **>SET-UP** and double-click, or keep pressing the **Tab** key until the frame around the word **>SET-UP** is flashing, and press **Enter**.

Figure 9.5 shows the menu that will appear. In the figure, the first item, **Phone number**, has already been chosen. This option lets you specify the phone number that the software will dial to access Prodigy Services. You can find your local Prodigy Services phone number in the Phone Book included in the PRODIGY Start-up Kit.

Figure 9.4
The initial PRODIGY Screen gives you the option to set up the PRODIGY software.

You select each of the menu items by pressing the **Tab** key until the associated > is highlighted, then pressing **Enter**. If you have a mouse, you can point to the > and double-click.

Like so many other packages, the PRODIGY software allows you to specify a primary and an alternate phone number—just in case one is busy or is otherwise inaccessible. The phone number we've entered here is the San Jose, CA number for 9600-bps service.

After entering the alternate phone number, you're ready to specify the COM port (the port that your modem is attached to). Figure 9.6 shows the choices that appear, with COM1 chosen, which is a port commonly used for external modems.

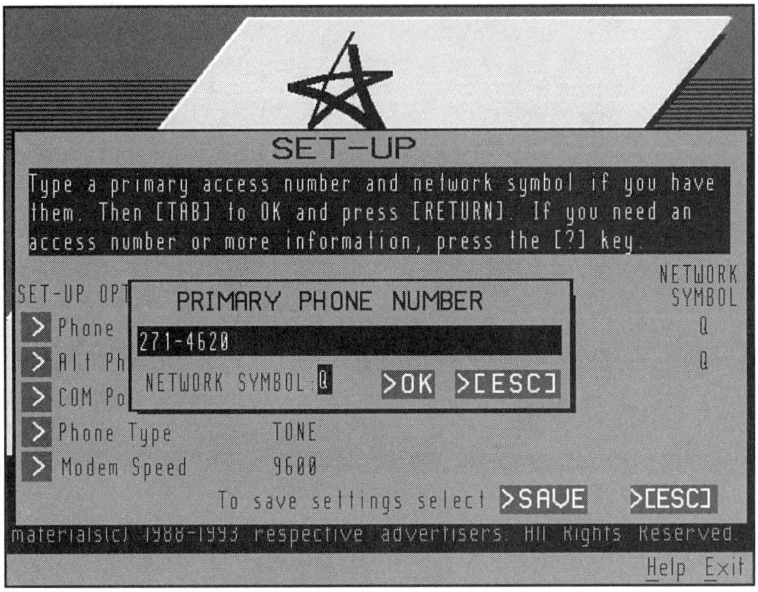

Figure 9.5
Entering the PRODIGY phone number that your PRODIGY software will dial.

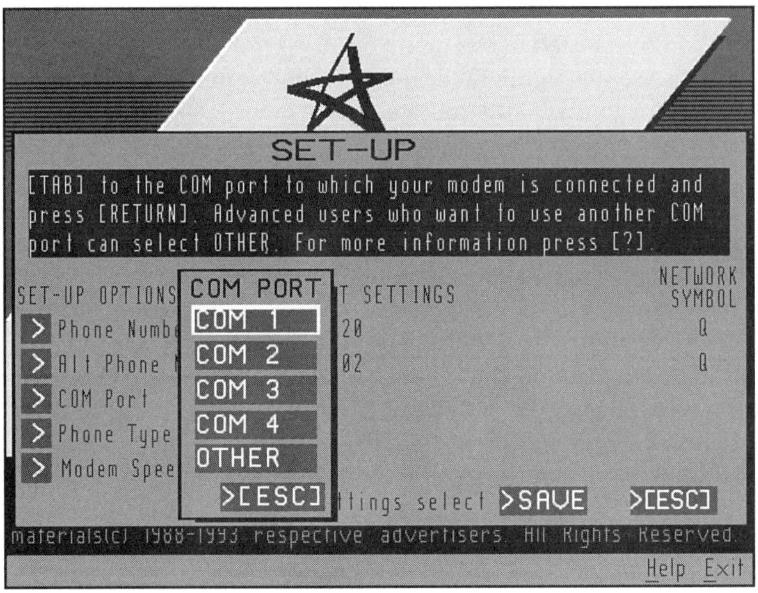

Figure 9.6
Setting up which COM port you'll use to go online with PRODIGY.

The OTHER option on the COM Port menu is just in case you have installed an internal modem (or an add-in serial communications adapter) that isn't addressed as COM1, COM2, COM3, or COM4. If you have such an arrangement, the manufacturer will have provided information on the use of a 4-digit *port address* (such as 03F8) and an *interrupt* (or *IRQ*) *number* (4, for example). PRODIGY makes two straightforward assumptions here—that you are what they call an "advanced user," and that you have a copy of the documentation for the internal modem.

If you bought a new phone in the last few years, you're probably familiar with the little Tone/Pulse switch on the side or the back. Phones are normally packaged with the switch already set for tone dialing. The PRODIGY software makes the same assumption, but lets you change it as shown in Figure 9.7. You should change it, of course, only if you are unfortunate enough to still be stuck with pulse-dialing service.

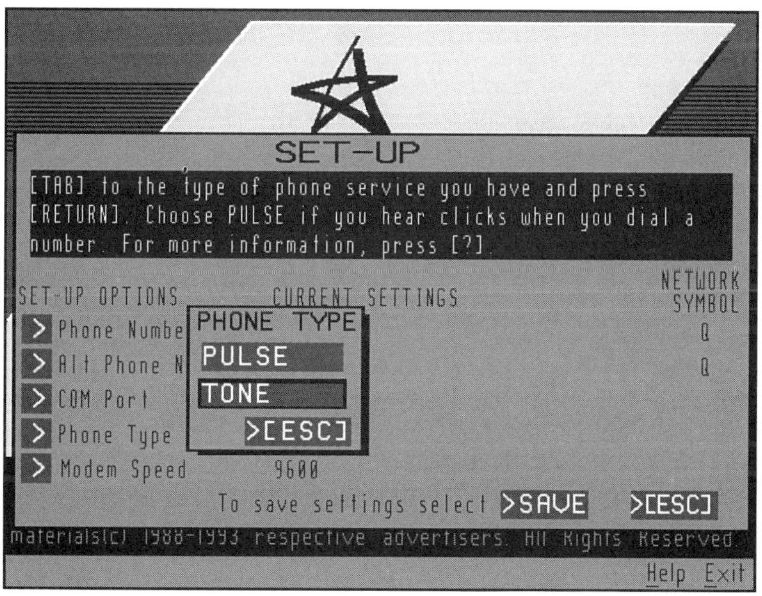

Figure 9.7
Change the phone type only if you need to.

Figure 9.8 shows the final Setup submenu. Note that if you choose 9600-bps operation, the Prodigy Service is polite enough to tell you that there will be a surcharge, and that you have to enroll for the faster service.

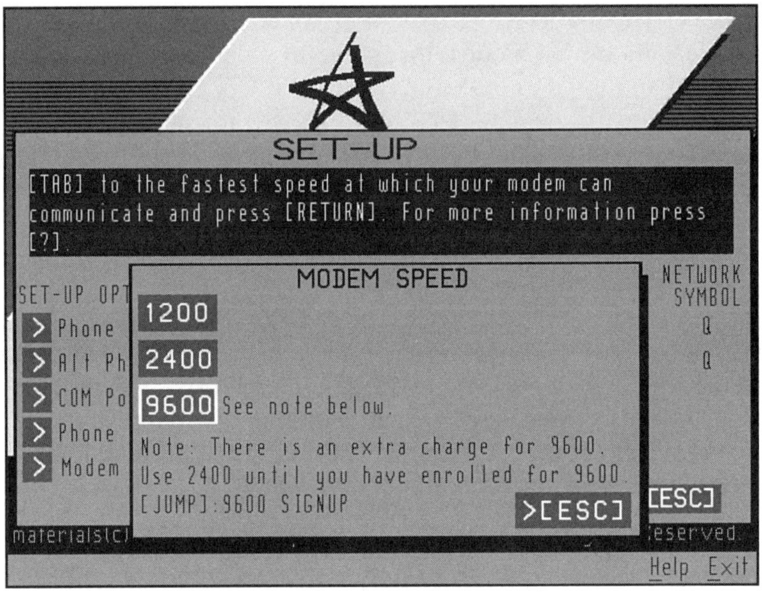

Figure 9.8
Specifying your modem's speed in PRODIGY.

What is absent from all of PRODIGY's setup menus is any kind of modem identification. This is because the PRODIGY software is designed to recognize what type of modem is connected to your specified COM port. Once it has done this, it will set it up with the correct command string.

COMMON SENSE

Don't panic if your copy of the PRODIGY software doesn't display the 9600-bps setup option. Just sign up for 2400-bps service. Then, any time you're online, you can JUMP to 9600 signup (press **J** for Jump, press **Enter**, type **9600 signup**, press **Enter** again) and follow the instructions that will then be displayed. PRODIGY will provide you with primary and alternate phone numbers for 9600-bps service in your area and will update your software automatically (see the end of Chapter 10) to add the 9600-bps option to the Modem Speed menu. Before you sign on next time, you can choose **>SET-UP** in order to enter the new phone numbers and select the 9600-bps option.

Establishing the Online Service Account

In their membership kits, CompuServe, the Prodigy Services Company, America Online, and others provide the means to establish your account.

Others, such as Delphi, provide an 800 number, a special, temporary user name, and a temporary password. Once you have entered the name and password, you are asked for name, address, phone number, a credit card number (so they can bill you), and other pertinent information. Once you have completed the registration process, you will receive a message to that effect.

Most of the neighborhood bulletin boards tell you that if you don't already have an account, you should enter your name. They then lead you through a question-and-answer session. If you are willing to provide the information they need, you will then be registered. Not all bulletin boards charge for their services. Those that do may charge only for some services. In any case, those that are not completely free may expect you to provide a credit card number.

PITFALL!

> There does not appear to be any evidence that neighborhood bulletin boards are anything but honest. However, unscrupulous people could conceivably set up a phony bulletin board in order to run a credit-card scam. So be cautious, but don't be scared off.

Exploring the Service

Once you're a member, subscriber, registered user, or whatever the online service or bulletin board cares to call you, it's time for exploration.

The larger services, such as PRODIGY and CompuServe, provide you with a clearly displayed menu item, inviting you to take a guided tour or to browse or explore the service, as soon as you sign on.

CompuServe's CIM (CompuServe Information Manager) software even allows you to choose a **Browse** option from the initial **Service** menu. When you choose it, CIM will make the connection, sign you on, and present you with the Browse menu.

Most of the local bulletin boards also provide a beginner's option on the first menu displayed after you sign on.

> **BEST BET**
> Whether you're new to online services in general, or just to a particular service, it usually pays to take the guided tour.

What Can Go Wrong, Go Wrong, Go Wrong?

"Welcome to the first fully-automated airline flight. All systems on this airplane have been rigorously tested, and each is backed up by two more identical systems. Please rest assured that nothing can possibly go wrong . . . go wrong . . . go wrong . . . go wrong . . ."

A number of things can, in fact, go wrong. Your software may encounter a situation for which it was not fully tested. The software used by the online service or bulletin board may be less than perfect. Your modem (or the one at the other end) may have a problem—and, of course, there could be problems with the phone line or service.

When something does go wrong, you will probably get an error message. If you do, don't panic. Just read it carefully, and follow whatever procedure it recommends. In many cases, the fault lies with the service, rather than something at your end. Chapter 19 of this book also offers some troubleshooting hints; these should help you identify and fix things that go wrong.

Ready for the Next Step?

You've reached the end of PART I of this book. PART II provides information that you may not have needed to get started, but which you may find useful now.

In the meantime, here are some of the things you have learned in this chapter.

- You need to install and configure either a generic communications software package, such as Windows Terminal, or service-specific software in order to communicate with an online service.

- You must pay attention to essential configuration information, COM port, online service phone number, modem commands, and so on. The nonessential options can be considered later.

- Service-specific software (PRODIGY's, for example) provides a limited number of choices. You don't have to specify modem commands, data bits, and so on.

- Online services provide you with full information, in their membership kits, on making the initial call and setting up an account. After all, they do want your business.

- Your first step, once you have established your online connection, should be to explore the online service.

- If something goes wrong, you can usually call an 800 number for assistance. This is especially true of PRODIGY and America Online, both of which assume that their members are not computer or communications experts.

CHAPTER 10

Exploring PRODIGY

Since it was introduced, the Prodigy Services Company's online service has enjoyed enormous market acceptance. As of March 1993, it had over two million members. Eight hundred thousand members sign on to the service during an average day, and up to forty thousand of them can be connected at the same time.

Starting PRODIGY and Signing On

In Chapter 9, we showed you how simple it is to configure the PRODIGY software. Starting it and signing on is as simple as typing **prodigy** at the DOS prompt, then following instructions on the screen to type your user ID (provided in the PRODIGY Start-Up Kit) and password. The password is also included in the kit, but any time you are online, you can change it to one you can remember more easily (just **JUMP PASSWORD** and follow the instructions).

You can also automate the sign-on procedure. Just **JUMP AUTOLOGON** when you're online, and follow the instructions. You will be asked to provide a nickname, which you will use whenever you want to use the service. Subsequently, you will type **prodigy**, followed by your nickname, at the DOS prompt. If you use KATHY as your nickname, you will type **prodigy kathy**.

WORD POWER

JUMP is one of several commands PRODIGY displays at the bottom of every screen (see Figure 10.1). You use it to go directly to a feature, a department, and so on. You can select it by pressing **F6**, or by pressing the **Tab** key until **JUMP** is highlighted and pressing **Enter**, or by pressing **J**, then **Enter**, or by using the mouse pointer and double-clicking on **JUMP**. Whichever way you do it, a new window will appear and you will be asked to enter your JUMPWORD (which may, in fact, be more than one word) and press **Enter**. You will also be provided with a short list of major departments (such as NEW MAIL) which you can select directly (**Tab**, **Tab**, highlight the selection, then press **Enter**, or click the mouse button). If you don't know what JUMPWORD to use, you can select an option which will display all available JUMPWORDs. You can also guess—which will get you a list of the JUMPWORDs closest to your guess. If the right one's there, you can select it.

PRODIGY's Start-Up Kit contains a booklet called "Getting Around." It explains how to use JUMP and all the other commands, including those intended for placing orders for products, for finding where to find things, for printing news items and other articles (or saving them to a file on your hard disk), and so on. The HELP command also stands ready to provide you with assistance.

PRODIGY lets you create a *Pathlist*, which you can customize to indicate the order in which you want to visit the various departments. You can move through your Pathlist, in sequence, by repeatedly pressing the **F4** key, or by using your mouse to click on the word **PATH**. You can go directly to a department in your Pathlist by pressing the **F3** key to reveal the list, then selecting any item. Any time you press F3 and display the Pathlist window (including the first time), you can select the **CHANGE** command. When you do, PRODIGY will guide you through the process of adding, deleting, or rearranging items.

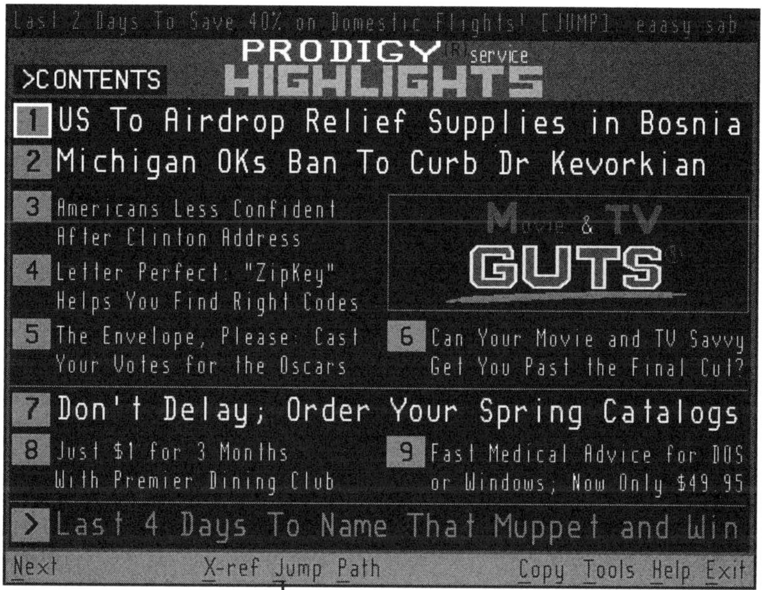

Figure 10.1
Highlights.

PRODIGY's Rates

PRODIGY charges by household. Each household can have a master account (whose number always ends with "A"). Additional accounts in a household have the same basic number, with B, C, D, or E as the ending letter. Billing is to the credit card of your choice. Effective July 1993, there are two payment plans.

PRODIGY Value Plan

The PRODIGY Value Plan is $14.95 per household per month, and includes 30 free messages (E-mail) per month, unlimited connect time for "Core" services, and two hours per month of free connect time for "Plus" features. These latter features include most bulletin boards, some stock market services (QUOTE TRACK, QUOTE CHECK, and Company News), and American Airlines' EAASY Sabre reservation system. E-mail messages in excess of the basic 30 per month are 25 cents each. There is no charge for incoming E-mail.

Once a household has used up its two-hour monthly allowance for the use of "Plus" features, the rate for additional hours is based on the following table.

Table 10.1 PRODIGY's overtime rates for "Plus" features.

Number of hours in excess of two	*Rate per hour for every hour*
per month	($)
Up to 3	4.80
3 to 6	4.20
Over 6	3.60

A quick calculation will show you that if your household uses 6 excess hours of connect time for "Plus" features, it might as well use up an extra hour ($25.20 in either case).

If you sign up before July 1993, you will have unlimited free connect time until June 30.

Alternate Plan

The Alternate Plan is $7.95 per household per month. It includes two free hours of connect time per month for "Core" services and "Plus" features combined. After that, the overtime rates are the same as for the Value Plan (but apply to all excess hours, whether the household member is using

"Core" services or "Plus" features). E-mail messages are 25 cents for every one sent (no 30-message allowance). PRODIGY recommends this plan to the occasional user.

Extra Services (Custom Choices)

PRODIGY offers several extra services. They are not subject to an hourly fee under the Value Plan, but they do have their own additional monthly fee. Table 10.2 lists those available as of April 1993.

Table 10.2 PRODIGY's Custom Choices

Service	*Price ($) and basis*
Strategic Investor	14.95/month/member
ZiffNet for PRODIGY	14.95/month/member, up to 3 hours
ZiffNet for PRODIGY	7.50/month/member, up to 1 hour
Baseball Manager-Classic	99.95/season/member
Baseball Manager-Lightning	49.95/season/member
FFL-Fantasy Football	99.95/season/member

With its announcement (April 1993) of hourly charges for "Plus" features, PRODIGY converted three of its former Custom Choices—PRODIGY Mystery Party, The Baby-sitters Club, and Sesame Street—to freebies. The PRODIGY Mystery Party has, however, a bulletin board that will be treated as a "Plus" feature.

To obtain information on a Custom Choice—and, if you wish, to sign up—all you need to do is select JUMP and enter the name of the Custom Choice that interests you. PRODIGY will take you, step by step, through all the details—and, if you decide to take the plunge, through the sign-up procedure. You will be billed on your credit card account, just as you are for the basic service.

Other Services on PRODIGY

PRODIGY Custom Choices are not the only extra services available. Other organizations offer their services on the PRODIGY network, and will typically bill the user's credit card or bank account for the use of those services. The most visible services are online banking and bill paying. BillPay USA offers online bill-paying services nationwide.

Table 10.3 lists banking institutions offering regional or local banking services, including bill paying. You can **JUMP** to **BILL PAYING** to select one, view their options, follow a demonstration, consider their rates, and sign on for their service—all online! To sign on, you are guided through a step-by-step procedure, which also explains costs, billing method, and so on. If you select an institution that doesn't serve your area, you will get a message to that effect, and will not be able to obtain any details of their service.

Table 10.3 Institutions offering online banking services on PRODIGY.

Institution	*Area Served*
Bank One	Columbus, Ohio
Boatmens	Missouri, Tennessee, Kansas, Illinois
Chemical Bank	New York
Chevy Chase	DC area
Comerica	Michigan
Delaware Trust	Delaware
First Interstate	Colorado
First PA Bank	Pennsylvania
Hamilton Bank	Pennsylvania
Mac Banking	Pennsylvania, Delaware, New Jersey
Meridian	Pennsylvania
Midlantic	New Jersey members
NBD	Michigan, Illinois, Ohio
NJNB Homebank	New Jersey members
Provident	Philadelphia

Institution	*Area Served*
Wells Fargo	California
Wilmington Trust	Delaware

Color, Pictures, and Ads, Ads, Ads

PRODIGY is like cable TV. You pay a flat rate for a good connection, then you either pay extra for premium services or accept commercial interruptions.

PRODIGY's commercials (or, rather, advertisements) don't interrupt you, but they do occupy part of the screen. If you want to see how much, look at the figures in this chapter. It is the revenue from such advertisements that allows PRODIGY to provide you with unlimited connect time.

Some advertisements are animated, either to a limited extent when they first appear, or to a much greater extent if you select the "LOOK" option that is displayed with each one. They can even play music on your computer's speaker, although few, if any, advertisers have chosen to exploit that yet.

You don't have to have a color monitor to appreciate PRODIGY. You can use monochrome graphics (even the old Hercules monochrome graphics adapter), but the colors, especially if you have a VGA, are really striking.

It's obvious that a major expense for PRODIGY is the payroll for the computer artists who create all the fancy illustrations and other graphics. Let's just go through a few of PRODIGY's many departments, and you'll see what we mean.

Getting the Daily Highlights

Each time you sign on, you'll get one of four Highlights screens—General, Arts, Sports, and Money. You can choose the one you want to see first, and

PRODIGY will remember your choice. In any case, you can use your PC's **Page Down** key to view the other three in turn. Figure 10.1 shows the main Highlights screen as it looked on the afternoon of February 25, 1993.

The Arts, Sports, and Money Highlights screens were very colorful on March 7, 1993. Note the offer of an online degree program. Chapter 12, which is about America Online, gives examples that show such programs are not unique to PRODIGY.

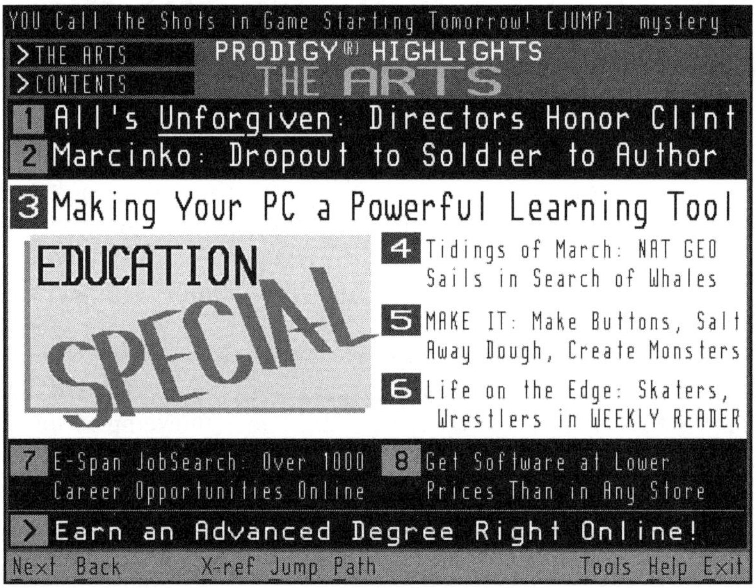

Figure 10.2
Arts Highlights.

You can use any of these screens as a starting point for the news or departments that interest you. In Figure 10.3, for example, you can select any of the numbered items (by entering the number and pressing the **Enter** key, or by double-clicking your mouse button on the item you want). Alternatively, you can select the general topics of Business and Investing. If you want the whole story, just select **Contents**.

The Sports Highlights screen not only offers you the sports news of the day, it also gives you a chance to find out more about one of PRODIGY's Custom Choices—the PRODIGY Network Golf Tour.

Figure 10.3
Money Highlights.

Figure 10.4
Sports Highlights.

Headline News

PRODIGY gives you several ways to reach the Headline News screen shown in Figure 10.5 (or any of a vast number of screens). You get there by selecting one of the two major news stories at the top of any Highlights screen, or by following your own Pathlist.

Figure 10.5
Headline News.

On this particular day, choosing item **9** from the initial Headline News screen displays not only a story, but also a map for those of us who might not know where Gila Bend is (see Figure 10.6). PRODIGY made extensive use of maps during the Gulf War.

Political Opinion—Jack Germond and Robert Novak

Who loves ya, baby? Or rather, whom do you love? If you're conservative, you'll probably love what Robert Novak has to say. If you're liberal, you'll prefer Jack Germond. If you're somewhere in-between, who knows? You might like both or neither. In any case, PRODIGY gives you a chance to find out, with a new column every day (although both of them have been known to let the same column do for two days).

Figure 10.7 shows what Jack Germond had to say on February 26, 1993. Note that, if you want to offer him your own comments, you can do so via PRODIGY's Close-Up Bulletin Board. (Sorry, Bob Novak, we called "heads for Jack," and the coin came up heads. At least your name is mentioned on this screen.)

Now you know where Gila Bend is. (But can you pronounce it?)

Figure 10.6
News Story.

The Inside Scoop—Karen Feld

Option 4 on the screen in Figure 10.7 takes you directly to Karen Feld's column (Figure 10.8), which she updates on Tuesdays and Fridays. If Novak and Germond are unable to arouse your interest in the political scene, may-be a little gossip will. Ms. Feld manages to be interesting without being malicious. Note the advertisement on the screen in Figure 10.8—PRODIGY's really doing the hard sell on 9600-bps service.

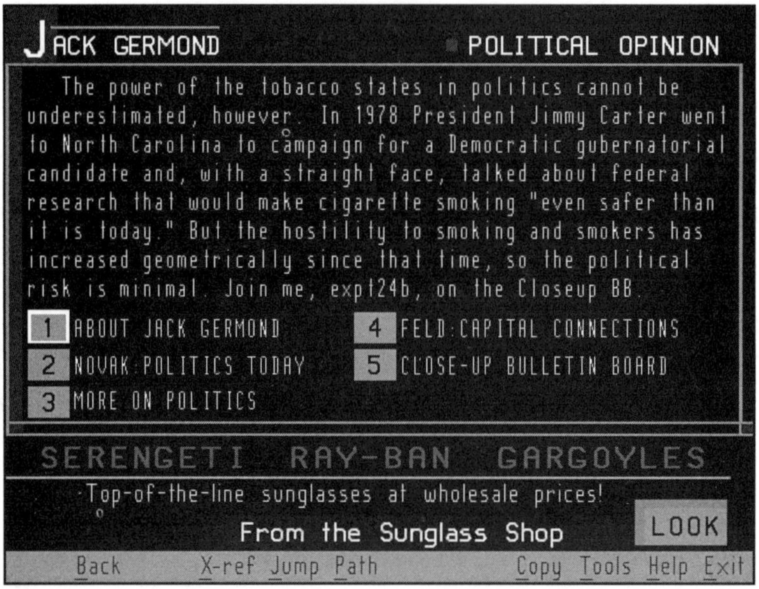

Figure 10.7
Getting a political opinion from Jack Germond.

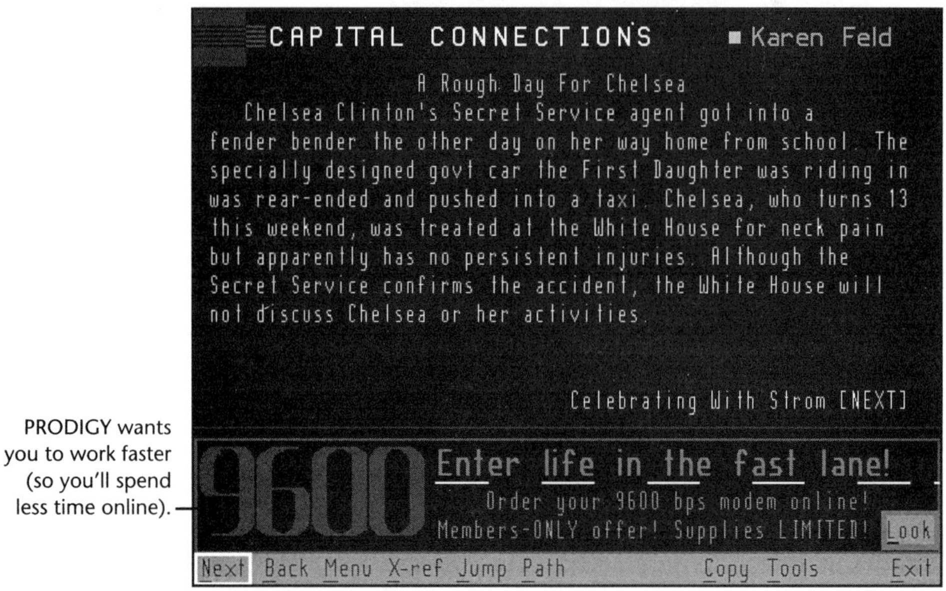

Figure 10.8
Karen Feld describes a rough day for Chelsea Clinton.

Weather

Everybody talks about the weather, but nobody does (or can do) anything about it. If you're about to leave on a trip, however, or if you want to reassure yourself that others are having worse weather than you are, you can **JUMP** to **WEATHER MAP** and look at the national weather map (see Figure 10.9). The Fronts map—which we don't show here—even shows you where the jet stream is, so you'll know in advance if your flight is going to be able to take advantage of it (or be slowed down by it).

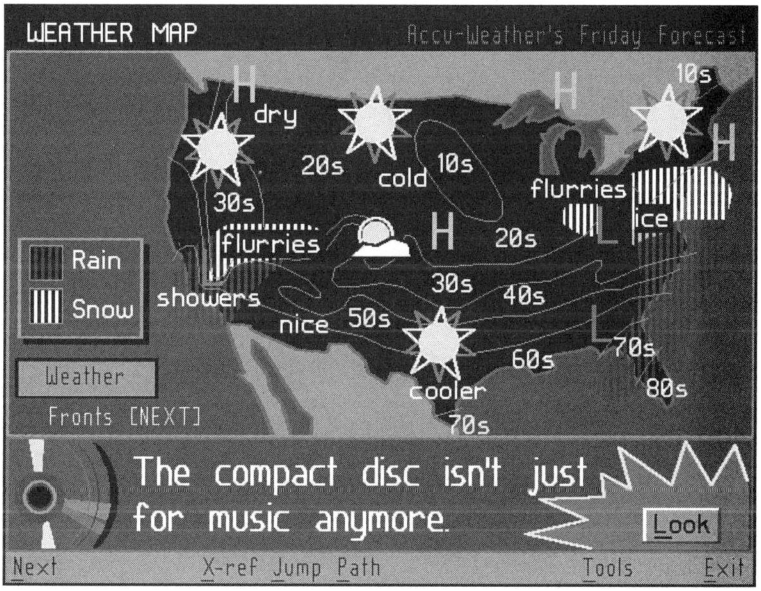

Figure 10.9
PRODIGY's national weather map.

If you want to see the day's forecast for your city, just select **JUMP**, type **W**, then a space, then the name of the city, then press **Enter**. Figure 10.10 shows a daily forecast for San Jose.

Going Shopping

Some people go to the mall to watch other people. Some go shopping. If you're in the latter category, you may be a browser, a buyer, or both. In any case, you'll probably want to visit the PRODIGY Mall. Just select **JUMP**, type **MALL**, and press **Enter**. Figure 10.11 shows the opening screen (complete with that pesky 9600-bps modem advertisement!). If you want to pass light bulbs on to your great-great-grandchildren, Item 1 is for you! If you prefer catalogs to whodunits, there's reading material you can order (lots of it).

Figure 10.10
City Forecast—San Jose, CA.

Close-Up Bulletin Board

Bulletin boards and forums are like nudist camps. They provide an opportunity for people to air their differences. Through its Close-Up bulletin board, PRODIGY offers the topic of the day (Today's Topic), and the opportunity to contribute your opinion on a number of topics continued from previous days. Figure 10.12 shows the Today's Topic screen; you can reach it by selecting **JUMP**, typing **CLOSE-UP**, and pressing **Enter** twice.

Heirloom-quality light bulbs. Leave them to your descendants.

Figure 10.11
PRODIGY Mall opening screen.

PITFALL!

PRODIGY has invited some First Amendment-related controversy by policing messages for vulgar language, personal insults, and so on. But the thing that is sure to get you in hot water with your fellow participants is posting messages that are all uppercase. Don't leave CAPS LOCK on—you'll be accused of shouting! (This applies to all bulletin boards and forums, not just PRODIGY's.)

Choosing one of the numbered topics displays a summary, and gives you a *jumpword* you can use if you want to participate (add your comments). As shown in Figure 10.13, if you don't want to discuss Bosnia with Novak, Germond, and many, many others, you can press the **Page Down** key to see what the opening proposition is on the President's economic plan.

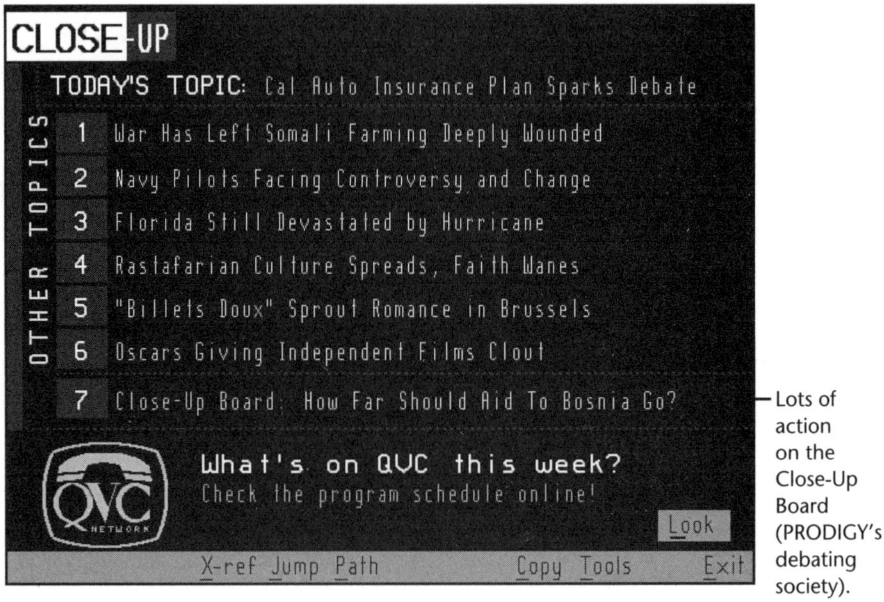

Figure 10.12
Today's Topic and other topics.

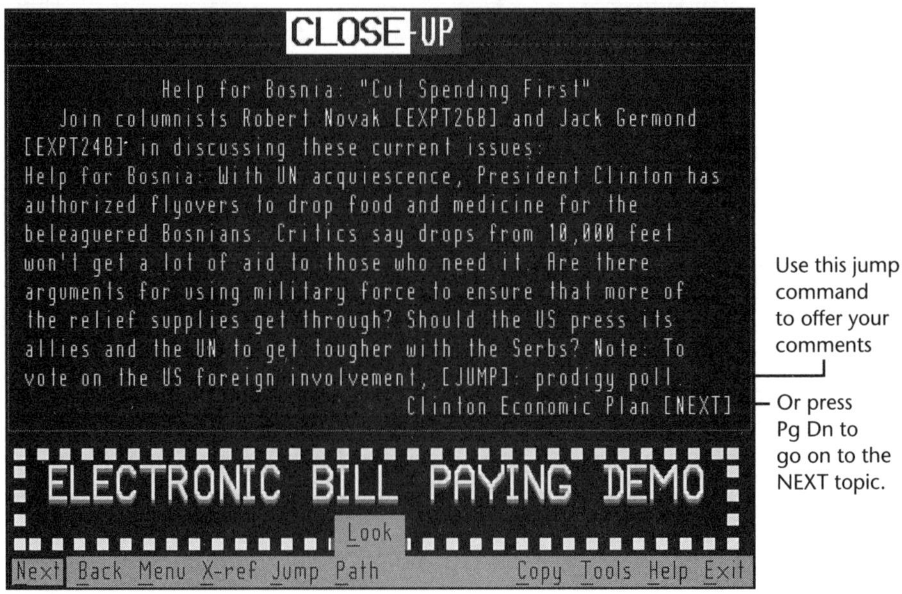

Figure 10.13
Invitation to participate.

Some opinions are vitriolic, some are off the wall, some are literary masterpieces, and some betray an aversion to correct spelling and punctuation. Contributions come from hawks, doves, pro-lifers, pro-choicers, liberals, conservatives, fundamentalists, nonbelievers, monarchists, abolitionists, reconstructionists, fascists, nihilists, pharmacists, Whigs, Tories, Luddites, . . . oops, mustn't get carried away. The one in Figure 10.14, reproduced with its author's permission, manages to combine good style with straightforward factual statements.

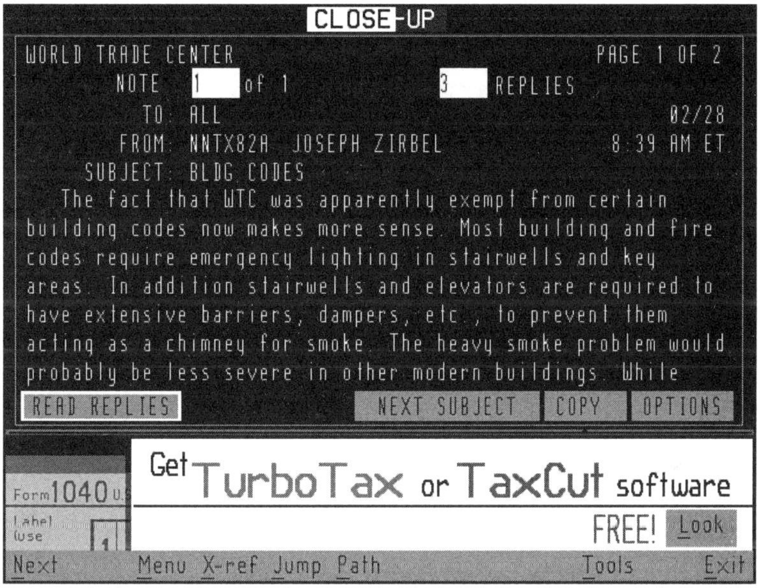

Figure 10.14
One of the less controversial Close-Up messages.

For the Kids

It's playtime. Finally, we get around to PRODIGY's kids' departments. There's lots of free stuff, and then there's Sesame Street (see Figure 10.15), a steal when it was $9.95 extra per month. Now the extra charge has been eliminated, so it's a *real* steal. Select **JUMP**, type **SESAME**, and press **Enter** to get there.

Figure 10.15
Welcome to Sesame Street.

And Lots More

This chapter can only scratch the surface of what's offered by PRODIGY and the other online services. Given that PRODIGY has over two million subscribers, there's a good chance that you know one or more of them. Ask around. And remember, if you try it and don't like it, you can always cancel your membership.

PRODIGY's software is constantly being updated—automatically and online! Any time there is an improvement or other modification to be distributed, there will be a slight delay when you sign on, while PRODIGY downloads and installs the changes in the PRODIGY directory on your disk drive. If you have just installed the PRODIGY software you bought at your local computer store, PRODIGY will recognize that it isn't the latest (how could it be?) and update it online.

There was a quite radical change in the software in the Summer of 1990, for which PRODIGY had to mail new distribution diskettes to all registered users. They haven't needed a mailing since then. PRODIGY has, however, announced new software for users of Microsoft Windows. It will be available in the second half of 1993.

CHAPTER 11

Going Online with CompuServe

Of the three or four best known online services, CompuServe is perhaps the one which best represents the "establishment." In 1988, CompuServe claimed to be the fastest growing online service. Since then, it has had to face PRODIGY's very aggressive marketing programs (and consequent rapid growth). The major benefit to CompuServe members has been the introduction of very competitive rates for monthly membership and connect time.

In this chapter, you will see what the dialog with CompuServe looks like when you go online with generic software (terminal emulation), and consider some alternative software choices. Then you will become acquainted with CompuServe's pricing structure. Finally, you'll tag along while we show you how easy it is to navigate American Airlines' EAASY Sabre reservation system, using the CompuServe Information Manager software.

Taking a Look Around

Your modem package will, very likely, contain a CompuServe booklet telling you how to sign up (and offering you a rebate on your first month's membership fees, or some similar deal). Your local computer dealer can also tell you how to subscribe, or you can call CompuServe's toll-free number: 800-848-8990. If you follow their instructions carefully, you'll find yourself with a membership number and a password—and, with your generic software, you'll be signed on. The instructions will tell you how to set up your software (data bits, stop bits, parity, and so on) so that it and CompuServe can end up talking the same language. (Oh, yes—those particular settings are **7** data bits, **one** stop bit, and **even** parity; see Chapter 9.)

Your initial dialog with CompuServe will look something like the one illustrated in Figure 11.1.

In the first line, the software is sending the setup commands to the modem, which the modem acknowledges in the second line. Then the software issues the dial command; a few seconds later, a connection is established (at 9600 bps, with error correction). Your next step is to press the **Enter** key. The 03SAA response tells you which "port" you're connected to (theirs, not yours). You are then prompted for the **Host Name**, to which you reply **CIS** (for CompuServe Information Service). Now you respond to the **User ID** and **Password** prompts, and—lo and behold—you are rewarded with a time stamp, confirmation of the type of membership for which you signed up, a copyright message, and CompuServe's "TOP Menu." (You can always return to this menu by typing **T** at any prompt and pressing **Enter**.)

Being anxious to find out how you can acquire a copy of the CompuServe Information Manager software, you ignore the menu and take the advice provided in *The Simple Modem Book*, which is to enter this command: **go doscim**. (Where and when is this advice provided? In this very paragraph, just now!) This evokes the response shown in Figure 11.2.

```
ATQ0V1E1*N6S0=0
OK
ATDT727-8113
CONNECT 9600 EC

 03SAA
Host Name:  CIS
User ID: 91119,9191
Password:

CompuServe Information Service
  02:07 PDT Sunday 18-Apr-93
      (Executive Option)
 Last access: 09:14 17-Apr-93

     Copyright (c) 1993
   CompuServe Incorporated
     All Rights Reserved

CompuServe                    TOP

 1 Access Basic Services
 2 Member Assistance (FREE)
 3 Communications/Bulletin Bds.
 4 News/Weather/Sports
 5 Travel
 6 The Electronic MALL/Shopping
 7 Money Matters/Markets
 8 Entertainment/Games
 9 Hobbies/Lifestyles/Education
10 Reference
11 Computers/Technology
12 Business/Other Interests

Enter choice number!go doscim
```

Figure 11.1
Signing on to CompuServe.

```
One moment please...

CompuServe               CIMSOFT

COMPUSERVE INFORMATION MANAGER (CIM) for DOS

 1 Introduction to CIM
 2 Standard Version Requirements/Features
    (Version 1.36)
 3 Advanced Version Requirements/Features
    (Version 2 - Requires HARD DISK)
 4 Download V1.36 Upgrade - No Charge
 5 Download Version 2 Upgrade
    ($10 Charge with $10 Usage Credit)
 6 Download German Version 2 Upgrade
    ($10 Charge with $10 Usage Credit)
 7 Order Software Kits
 8 DOS CIM Support Forum

Enter choice!7
```

Figure 11.2
Finding out about CompuServe Information Manager.

You decide to err on the side of caution and select number **7** from the menu. This, and your responses to some questions CompuServe will ask you, will bring you a copy of CIM in the mail, along with a manual. Had you chosen to download the version 2 upgrade, CompuServe's response would have included a message advising first-time users to order the kit instead. Besides, if you're using Windows Terminal to do this, your download time will be very long. That's because the only file transfer protocol Windows Terminal and CompuServe have in common is Kermit, which is quite slow.

While you're waiting for CIM, you'll have to content yourself with navigating through CompuServe with numbered menus and a repertoire of commands, including **go** (which you can use when you know the name of the department you want to go to), **find** (when you don't know where something is or what it's called), **m** (to redisplay the previous menu), **t** (to display the top menu), and **bye** (when it's time to sign off and get on with your life). These, plus many more commands, plus very comprehensive instructions, are included in the *CompuServe Information Services User Guide*, which is provided to all new members.

Software to Make CompuServe Life Easier

Four programs, including CIM, are currently available for downloading from CompuServe. As CIM is essentially free (because CompuServe will give you usage credits matching its price), we recommend that you obtain it first (as described); when you feel you're ready, use it to download one or more of the others (along with all the other stuff you'll now feel compelled to download). Any one of them will make life on CompuServe much simpler than generic software ever could. The programs are:

- *CompuServe Information Manager (CIM)*. CIM Version 2.1.2 is the latest release of this program.

- *CIM for Windows (WinCIM)* can also be ordered for shipment by mail. Its $15 price is offset by a CompuServe $15 usage credit. If you have Windows, an 80386SX or better, and two or more megabytes of memory, you may want to order this program rather than CIM. Figure 11.3 shows a CIM for Windows screen. Use **go wincim** to find out more.

> If you have an older PC or no hard disk drive, you'll need an earlier version (1.36).
>
> **PITFALL!**

- *TAPCIS (The Access Program for CompuServe Information Services)* is, like CIM, DOS-based. It is shareware, and (after a 90-day trial period) is subject to a $79 registration fee. Enter **go tapcis** for details.

- *OzCIS* is freeware, unless it is used for profit, in which case a $50 registration fee and 21-day trial period applies. The command **go ibmcomm** will help you learn more (but remember: the IBMCOMM Forum is an *extended service*; there is a connect-time charge).

BEST BET

> You can set up OzCIS to sign on, go to the forums you choose, locate all the new message subject headings since the last time you signed on, then sign off and disconnect. This is called the "First Automated Pass." It will then display the message headings, allowing you to mark those which interest you. It will then make its "Second Automated Pass" to retrieve and download all the messages whose subject headings you have marked, again disconnecting on completion. You can then read the messages off-line, creating replies if you wish.

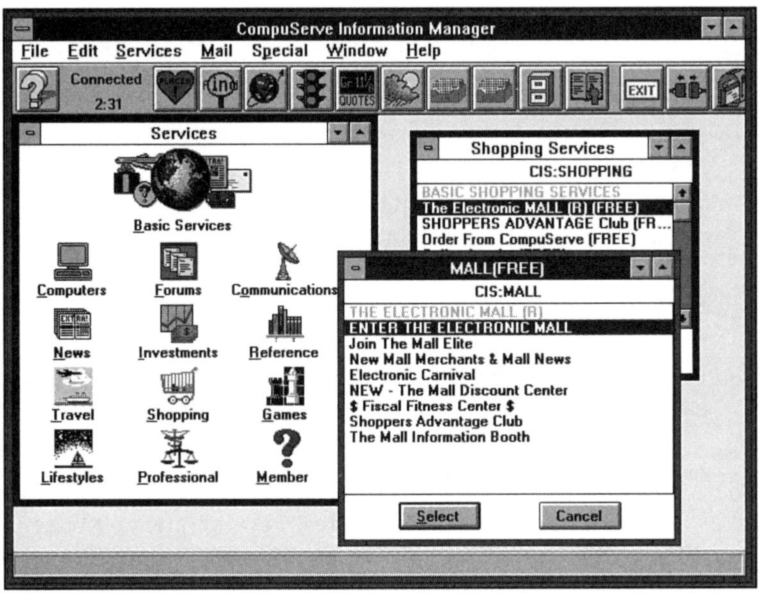

Figure 11.3

CIM for Windows (WinCIM) allows you to use CompuServe while you stay in the Windows environment.

Each package has its adherents and its detractors. All four packages provide some degree of automation, in some cases providing for automatic sign-on, automatic retrieval of user-specified information, and automatic sign-off.

Chapter 4 also provides some comments on these four programs.

What It's Going to Cost You

CompuServe offers two basic pricing plans, plus premium services, and a supplementary Executive Service Option.

Standard Pricing Plan

For $8.95 per month, you get unlimited connect time for a wide range of basic services. If you decide to use one of CompuServe's extended services, you will pay for connect time at a rate which depends on your modem's speed (see "Connect-Time Charges").

In addition to numerous forums, the basic services you can use for unlimited time periods include:

- NEWS, SPORTS, AND WEATHER
  Associated Press Online
      Hourly News Summaries,
      Sports, Entertainment, Business
      News, This Day In History
  Accu-Weather Maps/Reports
  National Weather Service
  UK News, sports and weather Clips

- REFERENCE LIBRARY
  Grolier's Academic American Encyclopedia
  Consumer Reports
  Peterson's College Guide
  HealthNet

- SHOPPING
  The Electronic Mall
  Shopper's Advantage (a discount shopping club)

- MONEY TALKS
  Basic Current Stock Quotes
  Issue/Symbol Reference
  Mortgage Calculator
  Fundwatch by Money Magazine

- GAMES & ENTERTAINMENT
  Roger Ebert's Movie Reviews
  Science Trivia Quiz
  The Grolier Whiz Quiz
  ShowBizQuiz
  CastleQuest
  Biorhythms
  BlackDragon
  Classic Adventure
  Enhanced Adventure
  Hangman

- COMMUNICATION EXCHANGE
  CompuServe Mail (see "Sharing a Little E-Mail")
  Classified Ads (reading only; placement costs extra)
  MAC, DOS, and Windows Support Forums for CIM
  Help Forum
  Navigator Support Forum
  Practice Forum
  Directory of Members
  Ask Customer Service

- TRAVEL AND LEISURE
  Travelshopper
  EAASY SABRE (airline, hotel, and rental car information and reservations)
  Zagat Restaurant Guide
  Department of State Advisories
  Visa Advisors
  Electronic Mall Merchants

Connect Time Charges

Under the Standard Pricing Plan, you must pay the following connect time charges, only when using extended services or premium services:

- At 300 bps, $6.00/hour.
- At 1200 or 2400 bps, $8.00/hour.
- At 9600 bps, $16.00/hour.

You can recognize extended services very easily. A plus sign (+) appears to the right of the name of the service.

Premium services are identified by a dollar sign (**$**), and always include a menu item which, if selected, allows you to view the basis on which you will be charged.

BEST BET

If your main use of extended services is for reading and creating messages on one or more of the special-interest forums, use the 9600-bps service, download all the messages you might be interested in, then either type **GO BASIC** or disconnect to read them. A V.32 modem, with V.42 *bis* data compression, will download the messages from four to twelve times as fast as a 2400-bps modem (V.22 *bis*, for example). That's a cost saving of 50 to over 80 percent. With CIM, TAPCIS, OzCIS, and so on, you make simple menu choices to mark messages for downloading, and to download them.

PITFALL!

When you go directly from an extended-services option to a basic-services option, you may continue to be charged for connect time. Avoid this by adding the option name to the word **BASIC** in the GO command (or GO menu item). For example **GO BASICEAASY** will take you from wherever you are to American Airlines' EAASY Sabre service, and will discontinue connect-time charges.

Alternative Pricing Plan

For $2.50 per month (called a Membership Support Fee), you get the right to connect to CompuServe. Once connected, you must pay for all connect time at the following rates:

- At 300 bps, $6.30/hour.
- At 1200 or 2400 bps, $12.80/hour.
- At 9600 bps, $22.80/hour.

PITFALL!

The alternative pricing plan is not a good value. Unless your activity is limited to a very small amount of E-mail (for example), with not much more than an hour of connect time per month, don't even consider it.

Executive Service Option

If you're going to spend any time at all on other services than basic (and you *will* if you develop an interest in any of the special-interest forums), go for this option—it's a no-brainer! All it does is subject your account to a minimum $10 monthly charge. Believe it or not, the $10 includes the $8.95 you pay for the standard pricing plan. If you spend more than four minutes per month on extended service options (at 9600 bps), your bill will exceed the $10 minimum.

If you actually use one of the executive services (Securities Screening, for example), you will be charged for connect time (as you are for extended services)—and, more than likely, will pay a surcharge (as you do for premium services). Those services for which you must have the executive service option are marked with an **E**. Those that are subject to a surcharge (most of them) also have a dollar sign (**$E**).

Remember, even if you never use it, the executive service option is essentially free (or effectively so, at $1.05 per month if you use only basic services and send only a moderate amount of E-mail).

Mailing Messages without Stamps: E-Mail

You can send and receive messages to and from other CompuServe users via CompuServe Mail (GO MAIL).

CompuServe charges nothing for incoming CompuServe Mail messages, except for messages received from Internet subscribers. Amazingly, the Standard Pricing Plan membership fee of $8.95 per month includes an electronic mail allowance of $9.00 (figure that out, you mathematical mavens). This is equivalent to about 60 three-page messages per month.

You can expect to pay a surcharge for fax, telex, and postal destinations, and for CongressGrams.

CompuServe mail also allows you to send messages to users of the Internet, MCI Mail, AT&T Mail 400, Western Union 400, Sprint Mail, or NetWare MHS. You can use the command GO MAILRATES to find out what any of these will cost you.

Try It! Using CIM to Navigate a CompuServe Service

We're going to look at how easy it is, with a program like CIM, to get into and use American Airlines' EAASY Sabre reservation system. If you already have your copy of CIM, you might want to follow along. If you haven't, come along anyway.

Loading CIM

If you've installed CIM properly (and with its fully-automated installation procedure, it's almost impossible not to), all you need to do is type **doscim** at the DOS prompt and press **Enter**.

Getting into EAASY Sabre

Now you've loaded it and gotten past the copyright message, you'll want to choose the **Go** option from the **Services** menu (see Figure 11.4). Enter the service name, **eaasy sabre** (although **eaasy** on its own would suffice).

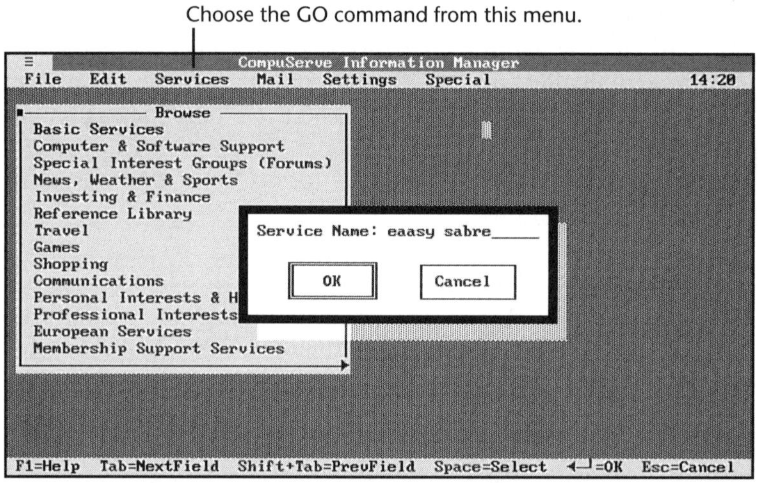

Figure 11.4
Entering the Service Name, "eaasy sabre."

Press the **Enter** key (or select **OK** and click the mouse). CIM takes the first step—dialing the local access number. The one shown in Figure 11.5 is the San Jose number for 9600-bps service. Connecting to CompuServe is just one of several messages you'll see as the connection is established and CIM signs you on.

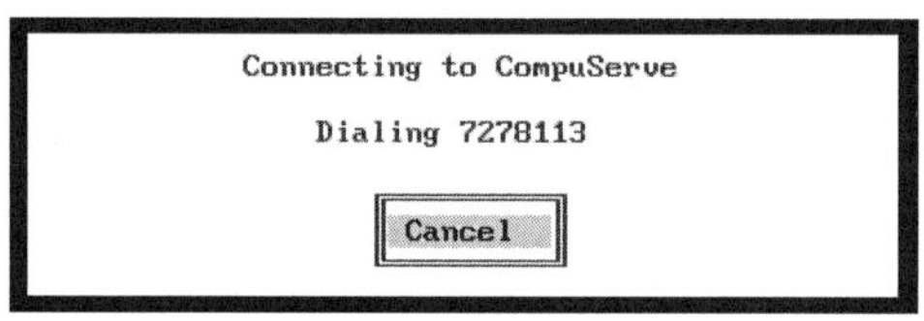

Figure 11.5
Connecting to CompuServe with CIM.

As soon as you're properly signed on, the EAASY SABRE menu appears (see Figure 11.6). At some point you'll want to select each of the first three menu items. For now, though, you're going to choose **Access EAASY SABRE (CIM)**.

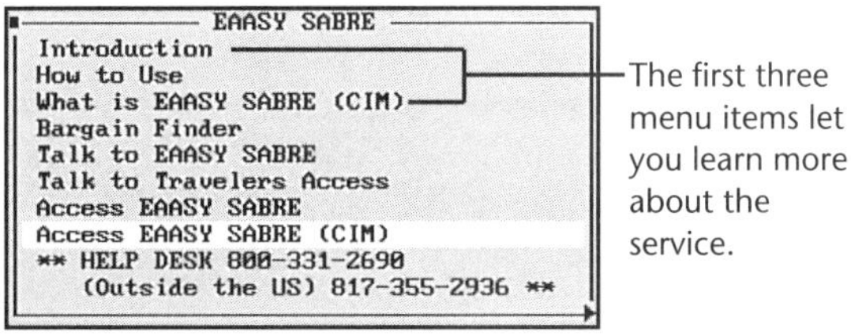

Figure 11.6
The EAASY SABRE menu.

If you had simply chosen Access EAASY SABRE, you would have seen CIM switch to terminal-emulation mode, where you would lose the advantage of using CIM.

Now comes the moment of truth. Do you have an American Airlines AAdvantage card? If not, you can apply for it right now (it's free), along with a no-charge EAASY Sabre account, just by selecting the **Apply** option. (If you were to actually do this, you would just answer a few questions, receive an AAdvantage number and make up your own password, and return to this menu.) Figure 11.7 shows a phony AAdvantage number and password (strictly for illustration!).

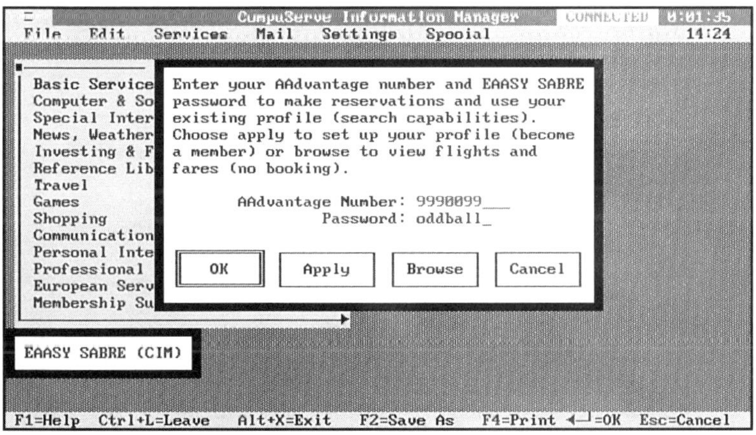

Figure 11.7
The initial EAASY SABRE screen.

Now you would enter a real number and password. A menu allowing you to choose Flights, Cars, Hotels, and so on appears; you should select **Flights**, causing the Destination window to appear (as shown in Figure 11.8).

Figure 11.8
The EAASY SABRE Destination menu.

If you entered what's shown in Figure 11.8, it means you want to fly from San Francisco to London on March 5, 1993, at or after 8:00 a.m. You'll be traveling alone. For San Francisco, you only entered **SFO**; EAASY Sabre filled in the rest. LON, by the way, covers Heathrow (LHR), Gatwick (LGW), Stanstead (STN), and Luton (LTN). If you didn't know the city code, you could have entered **London**, then EAASY Sabre would have asked you to choose between London, England (LON) and London, Ontario (YXU).

Your next logical choice is the **Flights** option, which provides you with the list of flights shown in Figure 11.9. Direct flights are first, followed by connections. Note, by the way, that the TWA "direct" flight involves a change of equipment (signified by CHG).

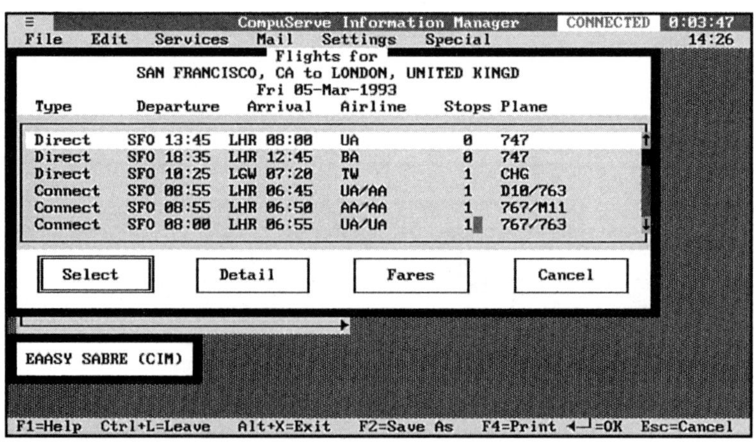

Figure 11.9
The first page of the San-Francisco-to-London flight schedule.

You could, for example, choose the United Airlines flight. No connections, and it puts you into Heathrow first thing in the morning. As Figure 11.10 shows, you have meal information, the type of aircraft, and the available fare classes.

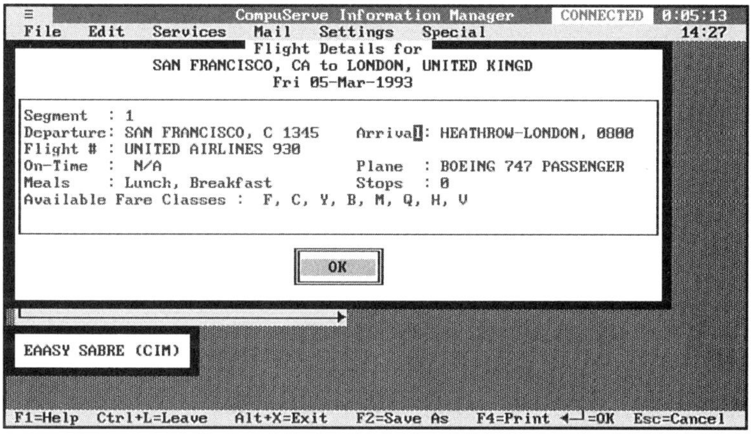

Figure 11.10
Flight details.

Pressing **Enter** (or clicking on **OK**) returns you to the Flight List (Figure 11.9). There you can choose **Fares**, which takes you to a menu where you can choose the class you want to fly—or, if you wish, be frugal and select the **BARGAIN FINDER** option, which lists fares starting with the lowest, regardless of class. You can select a fare, list the flights to which it applies, see what rules apply to a fare, and so on. Of course, if you're really traveling, you can go ahead and make a reservation, and ask for your ticket to be delivered to your house.

Getting Around Other CompuServe Services

A great many of the services available on CompuServe allow you to navigate through a series of menus and panels as easily as EAASY Sabre does.

Others will switch you into terminal-emulation mode, where you will still get menus. The only difference is that you must choose menu items by number, rather than selecting them directly (with the mouse, or by moving the highlight with the up and down arrow keys).

CHAPTER 12

America Online

America Online, another of the major online services, has been around since 1985. Until 1991, it was known only to Macintosh users. Since then, through aggressive marketing (most recently a Membership Kit and Tour Guide, including one month of free membership and 10 hours of free connect time), America Online has grown to over 200,000 members.

If your local bookstore or software store doesn't have a copy of the *Membership Kit and Tour Guide*, you might want to call America Online at 800-227-6364.

Setting Up America Online

America Online's setup is simplicity itself. You either install their DOS-based software and enter the **AOL** command from the DOS prompt, or you set up the Windows version and double-click on the **AOL** icon. You will then be led, step-by-step, through a registration process, in the course of which you will be connected to the America Online service. Subsequently, the only thing you have to remember is your password.

Just like the PRODIGY software, America Online's software determines, for itself, what kind of modem you are using. However, like CIM, it will subsequently let you modify things like modem commands.

Starting and Ending America Online

The Windows version of AOL is started just like any other Windows application—by double-clicking on its icon. For the DOS version, it's just a matter of typing AOL at the DOS command line.

In both cases, the initial AOL screen will appear. Figure 4.6 (Chapter 4) shows the Windows version. The DOS version's screen, also in the form of a window, is very similar, but lacks the button bar across the top. Your user ID, which AOL calls a *Screen Name*, is already in place; all you have to enter is your password. Figure 4.7, in Chapter 4, shows you what happens after you have done this. Again, the DOS version provides the same progress messages.

When you are ready to sign off, you can press **Alt-F4**, or move the mouse pointer to the top left corner of the screen and double-click (nothing unusual here). You will be asked for confirmation when you want to sign off, after which the window will remain; you can close it or sign on again. Once you close the DOS version's window, you will be returned to the DOS prompt.

Taking a Quick Tour of America Online

Next I'll show you just a few of America Online's departments to demonstrate how easy it is to navigate the service. Look at the America Online Windows screen in Figure 12.1. What is most striking about it is the number of icons—six on the Welcome menu and twenty arranged as a button bar across the top of the America Online window. By clicking on an icon, you can directly select most of the menu items that are otherwise accessible via the pull-down menus (Go To, Mail, Members, and so on). For example, you have four ways to select the screen shown in Figure 12.2. You can:

- Click on the NEW icon on the Welcome panel.
- Click on the NEW button on the button bar (it's there all the time).
- Select the **G**o To pull-down menu and click on the **N**ew Services option.
- Hold down the **Ctrl** key and press **3**.

You can edit the **G**o To menu to allow you to use the Ctrl key and a number key (0 to 9) to activate your own ten favorite features or services.

Pull-down menus

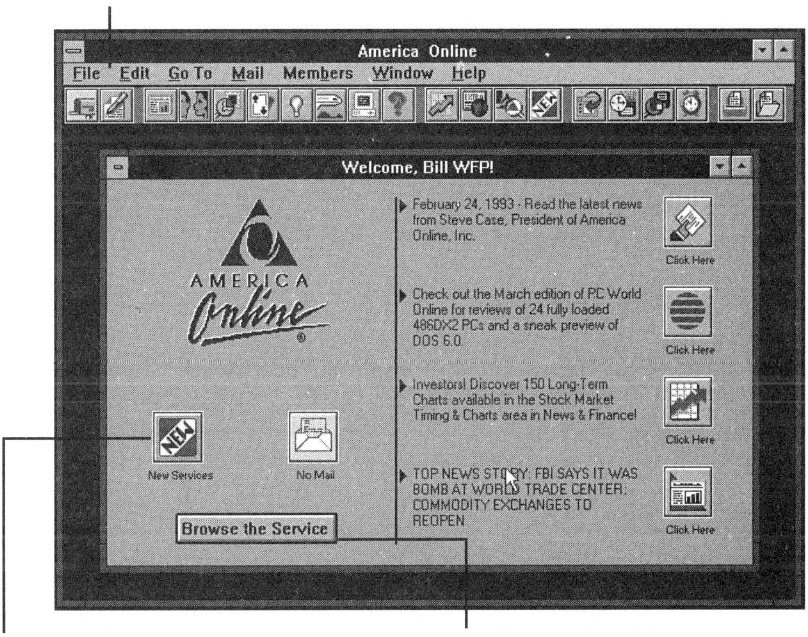

You can click on the NEW button in the Welcome window or the button bar.

Click on this button to get a menu of the departments.

Figure 12.1
America Online Welcome screen.

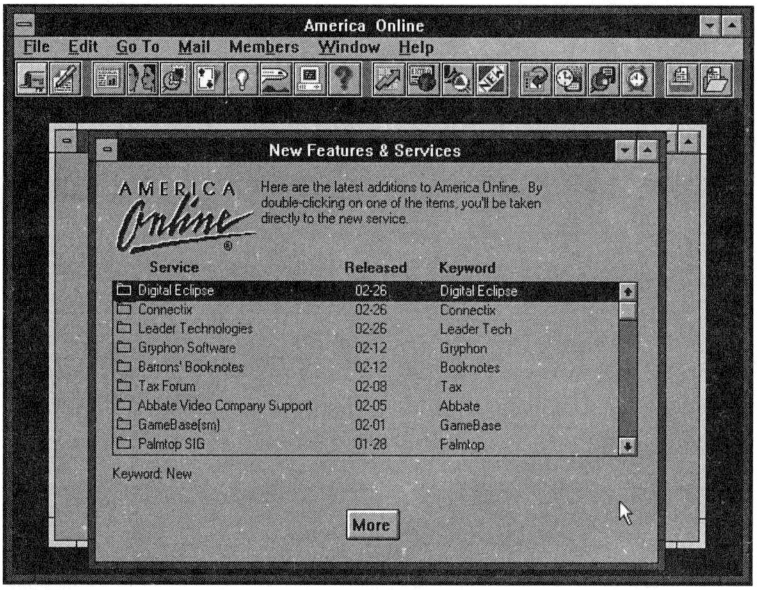

Figure 12.2
The New Features & Services screen.

Keeping You Informed

Like any successful supplier of goods or services, America Online tries to be innovative, and to add that little extra value to maintain your loyalty. The New Features & Services department (which you just learned how to find from the Welcome screen) is always available to update you on what America Online will be offering next. Figures 12.2 and 12.3 show eighteen new services and features that were introduced over a two-month period.

Checking Out the Departments

Choose the **Browse the Service** button from the **Welcome** menu (Figure 12.1), and you'll get the Departments menu (Figure 12.4). All eight of these icons are also on the button bar at the top. Once you're familiar with them, you can go straight to them.

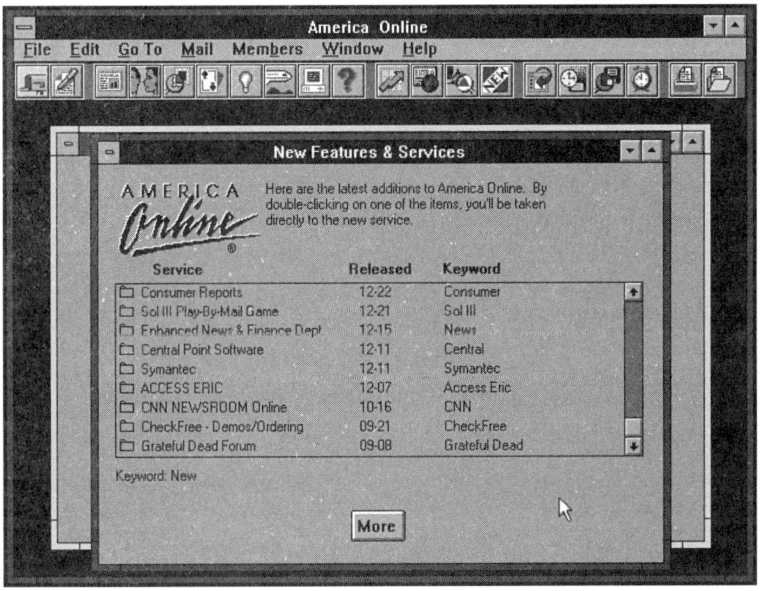

Figure 12.3
More New Features & Services.

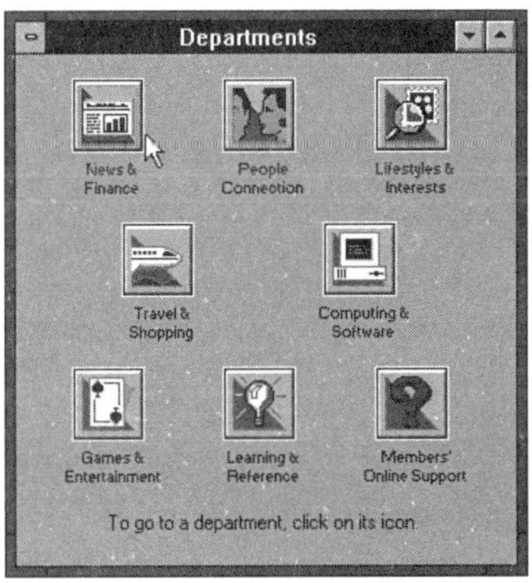

Figure 12.4
The Departments menu shows the various services you can access.

Using Keywords to Get to a Service

Do you want to get a degree online? This is one of the interesting options you'll discover if you choose **Learning and Reference** from the **Departments** menu. Doing so displays the screen shown in Figure 12.5. Note the keyword in the bottom left corner. This tells you that to go directly to this service, all you need to do is select the **Keyword** option from the **Go To** menu, and enter **EUN**. If you're handier with the keyboard than the mouse, all you need to do is press **Ctrl-K**, then enter **EUN**, then press the **Enter** key. Presto, you're there!

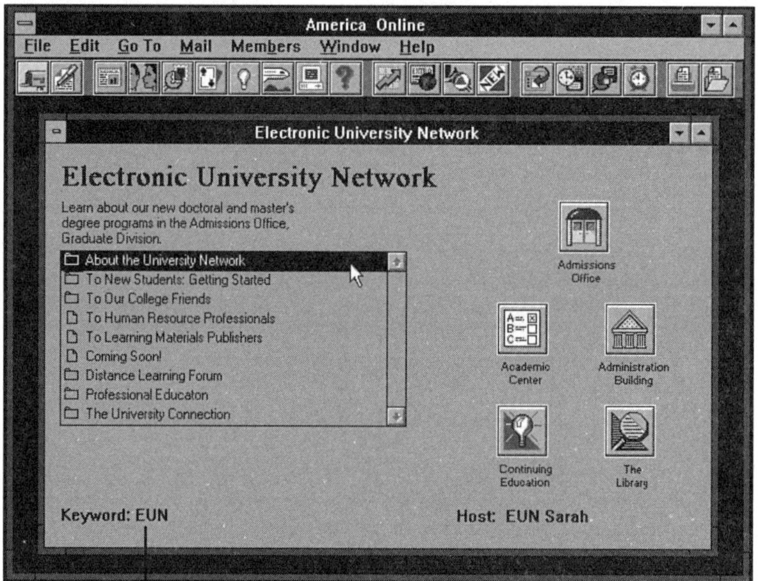

Use the Go To Keyword command, then enter this keyword, to go directly to a service.

Figure 12.5
The Electronic University Network.

As another example, you can use the **Go To Keyword** option to find out how soon America Online will let you connect at 9600 bits per second. Press **Ctrl-K** (or, if you're a rodent lover, use your mouse to select **Keyword** from the **Go To** menu) to display the screen shown in Figure 12.6. If you have trouble remembering the keyword, you can select the **Keyword List** button to view a list of the available keywords and where they'll take you. (In the DOS version, Keyword Help takes the place of the Keyword List.) The keyword for this example is easy to remember: 9600! Enter the keyword and press **Enter**, or click on **OK**. You'll see what's shown in Figure 12.7.

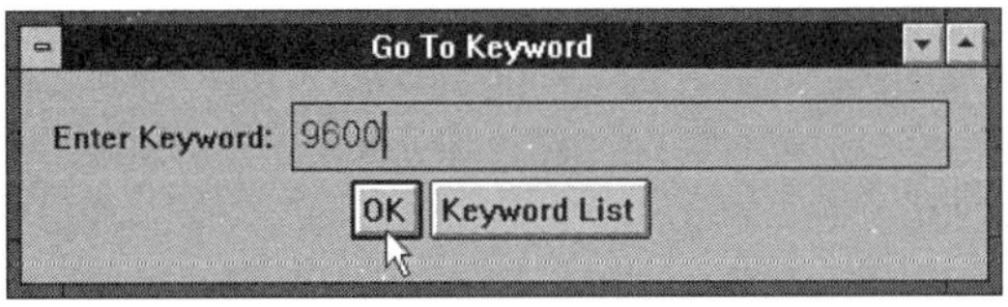

Figure 12.6
Go To Keyword.

Figure 12.7
9600-bps announcement.

Costs for 2400-bps Service

America Online's current phone numbers support modems operating at 300, 600, 1200, and 2400 bps. Connect-time charges are the same, whatever your modem's actual speed.

There is a membership fee of $9.95 per month (waived in the first month). There is no charge for the first two hours of connect time each month. After the first two hours, connect time is 10 cents per minute, unless you select the Members' Online Support department, which is free at all times. Currently, the rate for extra connect time (above five hours) is $6 per hour, or 10 cents per minute. As of July 1, 1993, extra connect time will cost $3.50 per hour.

If you're outside the 48 contiguous United States, you'll incur a 20-cents-per-minute communications surcharge, even when using the Members' Online Support department.

If you develop any software and get involved in sharing it with others, you might be interested in the 10-cents-per-minute credit America Online offers for the time it takes to upload something to one of the software libraries. Of course, you would have to upload an immense quantity of software for the credit to amount to very much.

America Online makes a point of emphasizing that, with three exceptions, it has no extra charges. The exceptions are their Quantum Space play-by-mail game (which charges per turn and per game), and Fax and Paper Mail, for which you'll incur per-addressee charges. Of course, if you order merchandise from one of the vendors on the service, you'll have to pay for that.

Plans for 9600-bps Service

America Online says it plans to offer 9600-bps service in the spring. By the time you read this, that should already have happened.

What Do You Get for Your Money?

Although America Online competes very well with PRODIGY and CompuServe, it has its own strengths and weaknesses.

If you enjoy discussions, you'll appreciate The People Connection, which allows you to get involved in real-time discussions with many other people on a variety of topics. Not all the conversations are as cryptic as the one shown in Figure 1.2 (Chapter 1). Although a number of the forums on

CompuServe sponsor conferences from time to time, they are quite tame compared to The People Connection, which is jumping all the time.

If you enjoy the more leisurely approach to discussions (that is, not in real time), you'll find a wide variety of forums in a number of categories, including games and entertainment, industry topics, learning and reference, lifestyles and interests, mutual interest, news and finance, special interests, and travel and shopping.

> **Parents beware!** Some of the People Connection discussions can be quite explicit. If you allow any of your children to sign on to America Online, you may want to monitor their activity.

One of America Online's strengths corresponds somewhat strangely to one of its weaknesses. In spite of a comprehensive library of graphics images, which you can download for display on your computer, you'll find you're unable to display them (or even the current weather maps) with the America Online software. PRODIGY displays weather maps as a matter of course, and the various CompuServe software packages (such as CIM) also allow you to display weather maps and other graphic images while online. On the positive side, America Online provides a DOS-based shareware program (which you can download and subsequently use) for viewing the downloaded graphics files.

25 Reasons to Love America Online

If you use the Go To menu to select the keyword **Unique**, America Online will provide you with 25 top reasons why you should love the service. Here are some of the things they say you can do:

- Download any of over 30,000 programs.
- Chat live and make friends in People Connection.
- Get help from experts and join computing forums.

- Access an international mail gateway.
- Play live-action games with members around the country.
- Get homework help and tutoring from professional teachers.
- Join others with your special interests to share information.
- Find small business information.
- Send faxes and paper mail without buying a fax modem.
- Attend college classes from home.
- Search the Wine & Dine Online wine database.
- Find out how to buy and sell property, refinance your home.
- Explore the National Geopgraphic Society Online.
- Create a stock portfolio with automated valuation.
- Tour the Library of Congress Online.
- Search for the latest news.

COMMON SENSE

Like the PRODIGY software, the America Online software is updated as needed—automatically and online! Check the revision number (*rev*) from time to time, by selecting the About America Online option from the **Help** menu. If it's changed from the last time, it means you've received an automatic update.

CHAPTER 13

Electronic Mail Options

From the point of view of the user (you), E-mail looks a lot like ordinary office mail. In a traditional office, you would create memos, letters, and so on, and place them in your out-basket. Every so often, someone (mail clerk, secretary, administrative assistant, or whoever) would collect the contents of the out-basket. Then, by the magic of interoffice mail, courier services, the U.S. Postal Service, or whatever, each item of mail would reach an addressee (possibly even the *correct* one). In the meantime, messages of various kinds (letters, memos, and so on), would be delivered at intervals to your in-basket; you would retrieve, read, and possibly answer them.

With the right software, electronic mail uses the same basic procedure, but different methods. The "right software," for the major online services, is the *service-specific software*:

- For CompuServe: CIM, CIM for Windows, TAPCIS, or OzCIS.

- For America Online: their proprietary DOS- or Windows-based software.
- For PRODIGY: their proprietary software.

Let's look at it, in its simplest form, step by step.

Creating Mail

Creating (or composing) mail does not normally require that you be online (PRODIGY is an exception). Usually, your software has a Mail menu, from which you can select a Create Mail item (America Online says Compose Mail). The software will request for the name and address of the addressee; this will include online service account number, user ID, or a similar unique identifier. After responding, you will enter a subject heading and the text of your message.

Sending Mail

When you complete your message, you will typically have two options—to send it immediately or to place it in an "out-basket" (on your hard disk). (You will also have a chance to place a copy in an existing electronic file folder, or to create a new folder and place it there). If you choose to send it immediately, your software will dial the online service number, sign you on, and send the message. If you place the message in the electronic out-basket, you can continue to create messages until you are ready to send them all. You will place the last message in the out-basket, and select an option to "send messages in out-basket." As before, the software will dial the online service, sign you on, and send all messages.

Receiving Mail

Your software can receive mail whenever you sign onto the online service, even at the time the software signs you on for the purpose of sending mail. When you sign on, the software at the other end checks an "in-basket" for all mail addressed to you. If there is any, it will probably allow you to choose between transferring it all to your in-basket or reading it (one message at a time). If you choose the former, you can sign off, then take your time going through the contents of the in-basket. Whichever you choose, you will have the option (as you did before sending a message) to place a copy in an electronic file folder.

Replying to Mail

Whenever you read mail from an electronic file folder, you will have an opportunity to create an immediate reply sometime after its receipt—even if you are have just retrieved the message. Because each message contains its author's name and address (and the subject heading), you have only to deal with the text of your reply. As with any other message, your software will let you send it immediately or place it in the out-basket.

Sending and Receiving Files with the Mail

Rather than creating and sending a message, you may wish to send a file. You will be able to either specify the file name as an attachment to a message, or to choose a Send File option from the Mail menu. In the latter case, you will be prompted for addressee information, just as you are when sending a message (plus, of course, the file name). When an online service has an incoming file for you, it will ask you what you want to call it (within the limitations DOS imposes for file names and paths), and will transfer it to your hard disk.

Electronic Mail and Terminal Emulation

You can, of course, send and receive mail with a terminal emulator like Windows Terminal. A number of online services, however, provide procedures that support terminal (or terminal emulator) connections; these include CompuServe, Delphi, GENie, and various bulletin boards. Though the procedures can be very tedious, many of the generic communications software packages include service-specific "scripts" for handling E-mail, especially incoming messages.

Beyond Creating, Sending, Receiving, and Replying

E-mail for personal computers goes beyond the facilities offered in the service-specific software for the large online services. If you work in an office with a *local area network* (*LAN*), you may be familiar with names like BeyondMail, cc:Mail, DaVinci eMail, and so on. The LAN may have an E-mail "gateway," through which it can exchange E-mail with other LANs, international mail networks, and/or employees' home-based personal computers. If your company wants you to connect to their E-mail system from your home, they will provide you with the necessary software and the instructions for its use. We'd like to tell you more here, but it's well beyond the scope of this book.

The "Other" E-mail Services

All of the major online services offer electronic mail; in many cases, they allow you to send mail to subscribers on other services.

Three of the possible groups to whom you can send mail are the subscribers to AT&T Mail, MCI Mail, and SprintMail. Although they may provide other services, these networks were established primarily to handle electronic mail.

Are Other E-Mail Services for You?

Before considering subscribing to AT&T Mail, MCI Mail, or SprintMail, you have to ask yourself, "How many people do I know who have an account with one of them? Am I likely to want to correspond with any of those people?"

The fact is that all of these systems are more suitable for business than for personal use. Their access (telephone) numbers also tend to be restricted to major metropolitan areas. If you have a business in such an area—and correspond regularly with other businesses that have accounts on AT&T

Mail, MCI Mail, or SprintMail—getting an account for yourself might make some sense. Otherwise you're better off with one of the general-purpose online services such as CompuServe, America Online, GEnie, Delphi, and so on.

One of MCI Mail's benefits in the past was that, with a 3-digit prefix, your MCI Mail account number was your Telex number. You could send messages to and receive messages from anyone in the world who had a Telex number. Now that most Telex machines have been replaced by fax machines, this capability doesn't seem so attractive anymore.

Rates for Other E-Mail Services

Currently, the three major E-mail carriers charge 5 cents per message unit. A *message unit* is defined as 1000 characters. Minimum monthly charges range from a low of $3 to a high of $20. Connect-time charges can range from $4 to $32 per hour, depending on the time of day and the type of connection (local or 800 number). All three carriers have procedures for sending mail to fax machines, telex machines, or postal destinations. These services all cost extra, with the amount varying by destination country. For faxes, companies can register their logo and copies of authorized signatures; a company's logo will be displayed on all of its E-mail faxes, with the authorized signatures appearing only on the signature owners' faxes.

BEST BET

> For a large company with a great deal of E-mail traffic, the minimum monthly charges are irrelevant. For the user of a home-based personal computer, wanting to send and receive E-mail for personal reasons or for a small, home-based business, the major E-mail carriers are too limited in scope (and, in any case, more expensive than the popular online services).

Extra Mail Options

In addition to its own internal E-mail system, each of the major online services (except PRODIGY) allows you to send mail to any of the others, either directly or through The Internet (a cooperative collection of networks serving government agencies, universities, and other institutions).

In addition to The Internet itself, cooperating networks include America Online, AppleLink, AT&T Mail, BITNET, CompuServe, CONNECT, Delphi, FidoNET, GEnie, MCI Mail, and SprintMail.

Surcharges for mail to other services vary. Messages from CompuServe members to MCI Mail subscribers cost $1 for every 7500 characters—slightly more than two and one-half times MCI's internal rate, but worthwhile if your mail volume to MCI Mail users is too small to justify a separate MCI Mail account. America Online makes a point of advertising its no-charge service for mail to subscribers to The Internet (plus several other services). CompuServe also imposes no surcharge for mail to The Internet—or, apparently, for mail destined for AT&T Mail, AT&T Easylink, or Sprint Mail users.

There is never a charge for mail received from another service.

Whichever service you subscribe to, you can easily find their instructions for sending mail to and receiving mail from other services . Most of them make replying to mail from other services as easy as replying to messages received via their own E-mail.

X.400 (What Is It?)

If you know someone with an "X.400 address," there's a good chance you can send mail to and receive mail from that address. In its X.400-Series Recommendations, CCITT (see Chapter 3) has defined a *message-handling system* (MHS) that many organizations have implemented, or are planning to implement. Again, online services such as CompuServe (which has an MHS "gateway") will provide you with the necessary information.

WORD POWER

Message Handling Systems (*MHS*) is the overall title of the CCITT X.400-Series Recommendations (X.400 to X.420). These define a global scheme for electronic mail, including mail that is to be delivered to postal or other noncomputer-based addresses. Over 600 pages of Recommendations define addressing schemes, protocols, procedures, terminology, and so on. The AT&T Mail, AT&T Easylink, and Sprint Mail services are designed to X.400 specifications. Also, Novell, Inc. provides NetWare MHS, whose users (whose computers are typically attached to NetWare local area networks) can send and receive mail, not only to and from one another, but also to and from CompuServe subscribers (via CompuServe's MHS Gateway) and any other users of X.400-based services.

CHAPTER 14

Bulletin Board Systems

Bulletin Board Systems, or *BBSes*, are small online services, usually operated by one person, a couple, or a special interest group.

They usually consist of one or more personal computers (or Macintoshes), with one or more modems and a great deal of disk storage, running special bulletin-board software such as Telix or WILDCAT! The software allows the owner to set up multiple message sections (by topic or interest, for example), libraries, user registration procedures, and so on.

Latter-Day Radio Hams?

Where they are run by one or two people, BBSes are, in some ways, today's equivalent of "ham" radio. They provide an opportunity to set up equipment, converse with others, and be a member of a special subculture. Just as ham operators feel compelled to upgrade their equipment constantly (and tell everyone else about it), so do BBS operators.

This particular BBS allows its users to engage in conferences with users on five other networks (and, therefore, with users of other BBSes that are also connected to one or more of the same networks). A simple screen picture won't even come close to showing you the number of conference topics accessible to any user. Table 14.1, below, copied from Terminal's screen with the Windows Copy and Paste functions, lists well over 200 conferences. To join a conference, all you have to do is enter its number in response to the message "Conference # to join (Enter)=." Using commands that will be displayed for you if you press H (for Help), you can read existing conference messages and contribute your own messages.

Table 14.1 Conferences available through The Higher Powered BBS.

The Higher Powered BBS Conference List

ILink	= -A	Intelec	= -C
Smartnet	= -B	Science Factor	= -E
Frednet	= -F		

SEE BULLETIN #10 FOR AN ALPHABETICAL LISTING

0 - Main Board	1 - HPB-ForSale	2 - Computer-News	3 - HPB-Fitness
4 - -Reserved-	5 - -Reserved-	6 - IL.Moderate-A	7 - IL.Admin-A
8 - Editors-A	9 - Emergency-A	10 - ILink.Info-A	11 - WordPerf-A
12 - Recovery-A	13 - SoundBoards-A	14 - HardDrives-A	15 - ForSale-A
16 - Chitchat-A	17 - Sports-A	18 - QmPro-A	19 - Robocomm-A
20 - Telecomm-A	21 - Hardware-A	22 - Shareware-A	23 - Graphics-A
24 - Politics-A	25 - Vegetarian-A	26 - Desktop-A	27 - Religion-A
28 - Virus-A	29 - 4Dos-A	30 - Windows-A	31 - Space-A
32 - Qmail-A	33 - Legal-A	34 - Investor-A	35 - TransAtl-A
36 - GayIssues-A	37 - Spreadsheet-A	38 - Multitaskng-A	39 - MMail.Rnet-A
40 - CountryMus-A	41 - Aviation-A	42 - Techtalk-A	43 - DOS-Tips-A
44 - Database-A	45 - Caveat-A	46 - Audio-A	47 - Writers-A
48 - PKWare-A	49 - QEdit-A	50 - HamRadio-A	51 - Beyond-A
52 - LapTop-A	53 - Amiga-A	54 - Scandinavia-A	55 - Philosophy-A
56 - Midi-A	57 - Gambling-A	58 - Ansi.Art-A	59 - Cars-A

60 - WordPlay-A	61 - RockRoll-A	62 - Sysops-A	63 - NetNews-A
64 - MacIntosh-A	65 - SmallBus-A	66 - LAN-A	67 - BasicProg-A
68 - Bookmarks-A	69 - SciFiction-A	70 - Collectors-A	71 - Psychology-A
72 - Gaming-A	3 - ZDCS-A	74 - Telix-A	75 - Guns-A
76 - Mods&Users-A	77 - DOS6-A	78 - ASM-A	79 - Paradox-A
80 - Jobs-A	81 - Mod.Meet-A	82 - PCTools-A	83 - -Reserved-
84 - BBS.Ads-A	85 - Doorway-A	86 - MS-Word-A	87 - Auto.Racing-A
88 - SharWareIss-A	89 - IL.RegAdm-A	90 - Ecology-A	91 - Espanol-A
92 - Pers.Financ-A	93 - Lantastic-A	94 - Win.App.Dev-A	95 - ProComm-A
96 - Pascal-A	97 - Build.a.PC-A	98 - Dr.Dos-A	99 - Soaps!-A
100 -Survivors-A	101 -Consulting-A	102 -MainFrame-A	103 -Comedy-A
104 -Unix-A	105 -Travel-A	106 -MobilStereo-A	107-HomeGarden-A
108-Programming-A	109 -France-A	110 -Pets-A	111-DownUnder-A
112 -Health-A	113 -Cuisine-A	114 -Coconuts-A	115 -History-A
116 -OLX-A	117 -Women-A	118 -C.Language-A	119 -USRobotics-A
120 -ClassiclMus-A	121 -Jazz-A	122 -Media-A	123 -Video-A
124 -AIDS-A	125 -Clipper-A	126 -OffLineRdr-A	127 -Astronomy-A
128 -Electronics-A	129 -OS.2-A	130 -CD-Rom-A	131 -Science-A
132 -StarTrek-A	133 -PhotoSig-A	134 -Micro.Help-A	135 --Reserved-
136 --Reserved-	137 --Reserved-	138 --Reserved-	139 --Reserved-
140 --Reserved-	141 --Reserved-	142 --Reserved-	143 --Reserved-
144 --Reserved-	145 --Reserved-	146 -Home&Garden-C	147 -Beatles-C
148 -NewUsers-C	149 -Smoking-C	150 -Deaf-HOH-C	151 -U.F.O.-C
152 -Legal-C	153 -GEOS-C	154 -NetAdmin-C	155 -NetStaff-C
156 -Markmail-C	157 -MartialArts-C	158 -Moderators-C	159 -Recovery-C
160 -C.Program-C	161 -Politics-C	162 -Pascal-C	163 -MIDI-C
164 -Motorcyc-C	165 -EMS.911-C	166 -Sysops-C	167 -ForSaleUsed-C
168 -ForSale-New-C	169 -Sports-C	170 -Comm-C	171 -Programr-C
172 -Medical-C	173 -Automotive-C	174 -BBS-Ads-C	175 -Wildcat-C
176 -DeskTopPub-C	177 -Database-C	178 -Literature-C	179 -PC.Security-C
180 -Shareware-C	181 -HamRadio-C	182 -PowerBasic-C	183 -Gay-C
184 -Mensa-C	185 -Spoilers-C	186 -Bicycles-C	187 -Windows-C

continues

Table 14.1 continued

188 -LAN-C	189 -OLX-SLMR-C	190 -Offline.Rdr-C	191 -Disabled-C
192 -OutDoorSprt-C	193 -Poetry-C	194 -MultiMedia-C	195 -Games-C
196 -Broadcastng-C	197 -CurrentEvnt-C	198 -Hobbies-C	199 -History-C
200 -Religion-C	201 -StarTrek-C	202 -Telemate-C	203 -Women-C
204 -PaintPgms-C	205 -Genealogy-C	206 -Survivor-C	207 -Teens-C
208 -Novell-C	209 --Reserved-	210 --Reserved-	211 --Reserved-
212 --Reserved-	213 --Reserved-	214 --Reserved-	215 --Reserved-
216 --Reserved-	217 --Reserved-	218 --Reserved-	219 --Reserved-
220 --Reserved-	221 -SciMathPrgm-E	222 -Geology-E	223 -Astronomy-E
224 -Chemistry-E	225 -ScienceProj-E	226 -H.P.-Users-E	227 -Herpatology-E
228 -EnvironEng-E	229 -ElectricEng-E	230 -Physics-E	231-QuantumPhys-E
232 -Mathematics-E	233 -Biology-E	234 -CivilEngin-E	235 -NASA-E
236 -Relativity-E	237 -SF.Admin-E	238 -MedicalSci-E	239 -Cognitive-E
240 --Reserved-	241 -FreeSpeech-F	242 -MensRights-F	243 -GayBashing-F
244 -Inequality-F	245 -DittoHeads-F	246 -LeftSpeak-F	247 -NPC-Humor-F
248-WtrMelonWko-F	249 -MushHeads-F	250 --Reserved-	251 -SN.Admin-B
252 -Music-B	253 -DESQview-B	254 -HardDrives-B	255 -IBM-B
256 -Comedy-B	257 -Pascal-B	258 -AmiPro-B	259 -TeenTalk-B
260-WordPerfect-B	261 -Telix-B	262 -Legal-B	263 -HSLink-B
264 -PKWare-B	265 -Windows-B	266 -BBS-Ads-B	267 -ForSale-B
268 -dBase-B	269 -Paradox-B	270 -OS.2-B	271 -Unix-B
272 -DESQview-X-B	273 -Borland-B	274 -Travel-B	275 -CD-Rom-B
276-C-Programng-B	277 -HamRadio-B	278 -Quicken-B	279 -Robocomm-B
280 -Auto-Fix-B	281 -Investments-B	282 -Lharc-B	283 -Telemate-B
284 -Hew-Packard-B	285 --Reserved-	286 --Reserved-	287 --Reserved-
288 --Reserved-	289 --Reserved-	290 --Reserved-	

Conference # to join (Enter)=

Bulletin boards like this one don't, in fact, maintain a full-time connection with other networks. Rather, they connect to them at least once a day to retrieve all messages added to each conference since the last connection, and to post their own members' messages to each of the conferences. On the screen shown in Figure 14.1, it says, "OVER 200 MILLION BYTES OF MESSAGES." That's no idle boast.

Special-Interest BBSes

These cater to an unimaginable range of interests, including antiques and collectibles, adult graphics, arts and crafts, games, gay and lesbian concerns, liberal and fundamentalist religion, classified advertising, horses, programming, astrology, astronomy, ad nauseam.

Company BBSes

Some companies establish specialized bulletin board systems to provide their customers with online support for their products. Subscribers can obtain up-to-date product information, download software fixes, and exchange messages with technical support people.

What Do They Cost?

Some BBSes, including company ones, are free. Some charge for downloaded software and other files (such as adult graphics), but charge nothing for E-mail. Those that charge may say so in their advertising, or may inform the new user at the time of registration. For those that charge, it costs nothing to find out how much or for what. Just connect to them with your communications software, and follow the instructions they provide.

Larger, private BBSes (such as Rusty and Edie's) will charge $30 and $50 for six months. Smaller BBSes ask for donations (such as $10 for six months), but these offer more limited services. Still others charge by the minute, and require you to pay in advance for a designated quantity of minutes.

BEST BET

> Look for brand-new BBSes; initially they will often offer free connect time to build their numbers of users and files. Also, check your local computer stores; they may offer free BBSes to their customers. You'll need to fill out a card in the store to join, and make occasional purchases to retain your membership.

The "Neighborhood" Boards and How to Find Them

Registration is usually quite easy. You dial the advertised number (using your general-purpose communications software), respond to a series of questions, provide a credit card number (for those that charge for various services), and suddenly you're a member of a brand-new community.

Go to your local supermarket, liquor store, or drug store. Outside, you'll almost certainly find a row of newspaper boxes, including ones for your local paper, *USA Today*, and a number of freebies, including real estate periodicals. More than likely, there will be one or more boxes containing *Micro Times*, *Computer Currents*, or something similar. These are the papers with the BBS advertisements. Other sources of these papers are computer stores and record stores (usually on a rack just inside the entrance).

To give you an idea of how many BBSes there are, the March 8, 1993 San Francisco Bay Area edition of *Micro Times* listed 64 for area code 408, 78 for area code 415, 65 for area code 510, 10 for area codes 209, 707, and 916 combined, and 9 with phone numbers in multiple area codes. Other metropolitan areas have similar concentrations.

What's a Sysop?

A *sysop* (a contraction of *system operator*) can be almost anybody. The owner of a BBS will usually identify himself or herself as a sysop, but may also enlist the help of others to be sysops for special-interest sections. On the large online services, such as CompuServe, all the forums have sysops, whose job it is to field questions and handle problems communicated to them by the many forum users. Where a forum is sponsored by a manufacturer (of a software product, for example), there may be paid sysops (employees of the manufacturer) and volunteer (unpaid) sysops.

Being a volunteer sysop can sometimes be a thankless calling. Users who are disgruntled about the product the forum exists to support will often vent their spleen on the volunteer sysops, even though they are responsible for neither the nature nor the quality of the product.

Forums

Forums are the bulletin boards of the large online services. Rather than dial a separate number for each bulletin board, you can sign on to the service and, using a "go to" option, join a forum. One service, PRODIGY, sticks with the bulletin-board terminology by consolidating all its forum activity under the "Close-Up Bulletin Board" banner. However, unlike the permanent forums of CompuServe and America Online, PRODIGY's bulletin board deals only with topics of current interest.

Live Conferences

These are difficult for many of the neighborhood bulletin boards, simply because of the limited number of simultaneous connections.

Live online conferences simulate a room. You can request a list of the people already in the "room," and you can enter it yourself. Once inside, you can see, on your screen, the messages making up the conversation. Each is identified by its originator's name or nickname (sometimes, CB-style, called a *handle*). To join in, all you need to do is type in your own two cents' worth.

Many of the CompuServe forums (or fora, if you want to show off your knowledge of Latin), sponsor an occasional conference. When they do, they usually notify forum users a week or two in advance.

With its People Connection, America Online makes live conversation a major feature of the service. Figure 1.2, in Chapter 1, illustrates one of the more esoteric conversations.

Are BBSes for You?

Should you go with the safe, sterile commercial service, or delve into the "underworld" of private BBSes? Here are some points to consider.

- Private BBSes are almost always cheaper than commercial services. Some are even free. The larger BBSes (usually not the free ones) have vast amounts of shareware and freeware to download, usually the same stuff as on the commercial systems.

- Almost all BBSes enforce limits on the amount of time you can spend online daily. Some BBSes enforce limits on the amount of software you can download in a 24-hour period. Some even have *upload/download ratios* that you must meet—that is, you must contribute (upload) a certain number of bytes for every kilobyte of data you download. Commercial services have no such limits.

- Private BBSes are usually run on PCs; they can experience periodic crashes, in which your messages may be lost (your ID may even be erased, forcing you to reregister.) Sometimes a BBS may be offline for days at a time while the owner repairs the system. Commercial systems, on the other hand, are usually run from mainframe computers; they will have backup processes in place so that the system rarely crashes or is offline.

PITFALL!

On commercial services, you are virtually assured that the software you download is legal for you to have. On some unscrupulous BBSes (known as *pirate boards*) there may be illegal copies of copyrighted programs that were intended to be commercial and not to be shared. These programs (often computer games) are usually "cracked" by someone who deciphers the code, modifies the program to remove its protection against copying, and posts

> the program illegally to show off. Some hackers make a hobby of this activity. Beware! Sometimes nasty viruses are hidden in these cracked games. Ask yourself: can you trust someone who would knowingly violate copyright laws to deliver a virus-free program?

- On commercial services, private E-mail is *private*, seen only by you and the recipient. On a BBS, the sysop (see Glossary) can look at anything he or she chooses, though most promise not to read private mail.

- With a subscription to a commercial service, you are subject to the rules and regulations of a board or committee of (one would expect) sane adults. On a private BBS, the sysop is godlike—he or she has the final say in all matters. If an unreasonable person is in that position, you're out of luck—with no recourse online.

- With a commercial system, you can exchange files and messages with many professionals in your field—for example, most professionals in the computer industry subscribe to CompuServe. With a BBS, if there are people you need to work with, usually you must ask them to join the same BBS; chances are slim that they will already belong.

- Unless you call a local BBS, long distance charges apply. Special third-party dial-in services can lessen some of the monetary impact, but these are often cumbersome and unreliable.

PITFALL!

> How do these third-party services work? They have "nodes" in most major cities. You dial into the local node, and tell the local service what number in what city you want to call. The service then connects you to a node in that city, which dials the phone on that end to call the board you want. The charges are about $1 to $3 per hour (much less than if you had dialed long distance yourself), but the line quality is not always good—and frequent accidental disconnections can occur.

CHAPTER 15

Transferring Files

If you've ever used the DOS COPY command or the Windows File Manager to copy a file, you're close to understanding what file transfer means. The difference is that, with file transfer, the original file and its copy are simply further apart at different ends of a communication link, to be specific. With an online service, you need a file transfer capability if you want to download files from the service or upload files to the service. You and a colleague may also want to send files to one another. Almost all communications software for personal computers includes a file transfer capability.

> **WORD POWER**
> *Uploading* and *downloading* are both file transfers. If you're transferring a file to an online service, BBS, or a fellow user's computer, you're uploading it. If you're transferring a file *from* an online service, BBS, or another computer (retrieving it), you're downloading it.

There are several advantages to sending and receiving files electronically. Here are just a few examples:

- You can download shareware or freeware programs from online services and bulletin board systems.
- Sending files (either by attaching them to an E-mail message as described in Chapter 13 or transferring them directly to another user's computer) is less expensive than using an overnight mail service.
- File transfer is almost instantaneous, making it faster than overnight mail.
- File transfer is ideal when you want to edit a document. The recipient can make changes, then upload a copy of the edited file to the original sender.
- Often, you can download information that's more current than the information in published resources, such as magazines, without having to make a trip to the library to make photocopies of your resources.

When you *transfer* a file, you're basically sending or receiving a copy of it over the phone line, to or from the source your modem is connected with (another computer, online service, or BBS).

WORD POWER

Downloading is retrieving a file to your computer, receiving it from an online service, BBS, or another computer.

Uploading is transferring (sending) a file from your computer to an online service, BBS, or another computer.

What You Need to Know About File Types

In the world of file transfers, there are two types of file—text files and binary files.

Text files contain displayable (or printable) characters only. They are usually the displayable characters of 7-bit ASCII. They may sometimes be all but the lowest 32 characters of 8-bit ASCII. When a text file is being downloaded, the software will, typically, display each line on the screen. What appears on the screen will also be written to the file, on disk, you have specified.

If a PC file has the extension .ASC, .TXT, or .BAT, it may qualify as a text file—*but*—see the PITFALL TIP, next!

> You cannot always rely on either common sense or the PC-based definition of a text file. CompuServe's Users Guide specifically says that the contents of a text file are limited to 7-bit ASCII characters, which excludes the accented characters used by most of the European languages.

PITFALL!

Binary files can contain anything in any character position. Program (.EXE, .COM, .DLL, etc.), word processor (.DOC, .DOT, etc.), graphics (.AI, .BMP, .CDR, .EPS, .GIF, .IMG, .PCT, .PCX, .WMF, etc.), spreadsheet (.WRK, etc.), and most other files are binary.

One of the most significant advantages of binary file transfer protocols is that they always include error checking. So, it's always safer to use a binary file transfer, regardless of file type.

If binary transfers are always safer, why would you ever want to consider a text file transfer? One example comes to mind. Suppose a friend or colleague has a very large text file that might interest you. Using Windows

Terminal (for example), you and your friend might want to start a transfer of the file. As it arrives, you will see each line on the screen. If you decide you're not interested in the file or, at least, in any more of it than you have already received, you can terminate the transfer. You might terminate it even if you are interested in having a copy, just so you can then restart the transfer using one of the error-free binary file transfer protocols.

Text files contain displayable (or printable) characters only. They are usually the displayable characters of 7-bit ASCII. They may sometimes be all but the lowest 32 characters of 8-bit ASCII. If a PC file has the extension .TXT or .BAT, it probably qualifies as a text file.

Binary files can contain anything in any character position, including instructions from the program that created the file. It's best to treat any files other than those with .TXT and .BAT extensions (and READ.ME) files as binary. For example, most word processing files have formatting instructions (such as those specifying different fonts and margins); that makes these files binary files.

For file-transfer purposes, a text file can be treated as a binary file, but a binary file cannot be treated as a text file.

File Transfer Protocols

Just as there are protocols governing the discourse between diplomats, there are protocols governing the conversation two computers must have in order to transfer a file, without error and without alteration, from one to the other.

File transfer protocols define the number of characters to be sent at one time (block size), how transmission errors are to be recognized, how the receiving end is to notify the transmitting end that it has detected an error, how error recovery is to be managed (including when to give up retrying the transfer of a block), how file names are to be identified (for batch transfers of multiple files), how file transfer completion (end of file or, for batches, end of last file) is to be signified, and so on.

WORD POWER
A *block* is a succession of characters, sent without pause. File transfer blocks may be from 96 to 1024 characters (or bytes) in length (or even longer). They contain a checksum (see "The XMODEM Family," later in this chapter), which allows the receiving software to confirm that the block is free of transmission errors. Based on the checksum, the receiving software must notify the sending software either that the block is good or that it needs to be retransmitted. If the characters of a file were not assembled into blocks, the receiving software would have to acknowledge every character individually—resulting in a very slow rate of transfer.

Many file-transfer protocols have been developed over the years. Some are inherently faster than others; some are more robust (they detect errors more reliably); some are simple, and some are complex. Fortunately, where they are complex, the complexity is hidden from the user. They're all equally easy to use. Usually, they're easy to select. This is because most online services present you with a limited choice (or no choice). In any case, your choice will be limited by the protocols supported by your communications software. If your software is service-specific, it will almost certainly select the protocol for you (and won't even have the good grace to tell you which one it selected—Rats!).

Many schemes have been developed to transfer files, without alteration, over a communications link; these are known as *protocols*. In all cases, they send the file as a series of short blocks, until the entire transfer is complete. As each block is received, it is checked for errors. If there are any, the receiving software asks the sending software to send the block again. Let's look at some of the available choices.

The XMODEM Family

The original XMODEM protocol was designed in 1977 by a computer enthusiast named Ward Christensen. With XMODEM, files are sent in 128-byte blocks, with one byte added for checking purposes. The extra byte, called a *checksum*, is the eight low-order bits of the sum of the 128 bytes. The receiving software also adds the bytes together, and compares the result with the checksum. If the checksum transmitted does not equal the one calculated by the receiver, the receiver requests retransmission of the block.

The XMODEM-CRC protocol replaces the checksum with something called a *cyclic redundancy check*. It serves the same purpose as a checksum but is more reliable (for reasons, based on the way it is calculated, too complex to explain here).

For today's high-speed modems and reliable communication links, 128 bytes is a less-than-ideal block length. A newer version of the protocol, XMODEM-1K, increases the block length to 1024 bytes.

The YMODEM Family

YMODEM protocols correspond to XMODEM protocols, but can adapt to noisy line conditions by switching automatically from 1024-byte to 128-byte blocks. Shorter blocks are less likely to encounter line problems during transmission (less time for it to happen), and can be retransmitted more quickly.

YMODEM protocols also use a special *header block* at the beginning of the transfer, containing the file name. This simplifies the task of transferring files in batches of two or more.

The ZMODEM Family

ZMODEM protocols add features which improve transfer performance and error recovery. One of these is the capability of resuming an interrupted transfer.

Kermit (Not the Frog)

Kermit was designed for file transfers between PCs and host computers. It is seldom used by online services although CompuServe supports it). However, Microsoft's Windows Terminal program supports Kermit (along with XMODEM-CRC).

Proprietary

The advantage of using service-specific software is that you need not be concerned about protocols. The software takes care of everything; online services may use their own protocols. CompuServe, for example, uses one of its own protocols (B, B Plus or Quick-B) when transferring files to or from a personal computer that uses CIM, TAPCIS, or OzCIS software.

There are other file-transfer protocols, such as the very efficient HyperProtocol, provided with Hilgraeve's HyperAccess communications software. As with any protocol, the proprietary ones must be implemented at both ends of the communications link. This tends to limit the use of HyperProtocol (and others like it) to direct links between personal computers.

Compressing and Decompressing Files

HyperProtocol is also an example of a protocol that can compress and decompress files "on the fly." For most protocols, it pays to compress files prior to transmission and decompress them after receipt.

PKZIP and PKUNZIP

These are shareware utilities for data compression and decompression, subject to a registration fee that you are honor-bound to pay. For many files, PKZIP can achieve a 5 to 1 compression. If you download a file with an extension of .ZIP, you will have to use PKUNZIP to decompress it. Some compressed files are *self-extracting* (they contain the program that decompresses them), and have an extension of .EXE.

You have two good reasons to get your own copy of PKZIP and PKUNZIP. First, you will save file transfer time and, if you're paying for connect time, money. Second, many of the files available for downloading have already been compressed with PKZIP; without PKUNZIP, you can't use them. Most bulletin board systems and many online forum libraries contain a self-extracting file, PKZ110.EXE, containing copies of PKZIP.EXE, PKUNZIP.EXE, and PKSFX.EXE, plus full documentation. (PKSFX converts a compressed file, with an extension of .ZIP, to a self-extracting file, with an extension of .EXE.) If you subscribe to CompuServe, you can use the command **go pkware** to get you to the PC Vendor C Forum, where you'll find PKZ110.EXE in Library 11 (PKWare). You may already have your own copy, and it's very possible you know someone who has a copy. As a last desperate measure, you may want to call PKWare, Inc. at (414) 352-3670. PKWare's documentation tells you about the benefits of paying to register their product.

PITFALL!

PKZ110.EXE contains version 1.1 of PKZIP and PKUNZIP. This is the most widely used version, but PKUNZIP 1.1 cannot be used to extract files that were compressed with the current version of PKZIP (2.04). You may wish to download that version, too. However, most people who upload files to CompuServe continue to use Version 1.1.

Selecting the Right Protocol

Customarily you will be faced with a limited choice of protocols, or no choice. If you are using Windows Terminal, your choice is among *text transfer* (not really a protocol), XMODEM-CRC, and Kermit. For PC-to-PC transfers, you should choose XMODEM-CRC. For PC-to-host or host-to-PC transfers, use Kermit only if you have to (it's relatively slow). If the host supports XMODEM-CRC, use it.

To transfer files using other general purpose software, read the manual and follow its recommendations.

The File Transfer Process

The documentation provided when you subscribe to an online service, plus your service-specific or generic communications software user guide will tell you how to go about uploading and downloading files. For generic software, the process doesn't vary much from one online service or BBS to another.

Getting to the library to which you want to upload a file from or which you want to download a file is, typically, an exercise in choosing items from a succession of menus. Once you have reached the library and, if necessary, browsed through it (also with the assistance of menus), you can select downloading from a menu. There may be several menu items covering a range of protocols, or you may select the protocol on a subsequent menu. You will be asked to name the file you wish to transfer and, having provided it, you will be asked to start the transfer at your end. You will do this by selecting your software's file transfer **Receive** option and specifying the protocol (matching the one you already selected) and the file name (including path) to which you want the transfer to be made. At this point, the download process will start, and normally will run to completion. The process for uploading is very similar. Choose your software's **Send** option and enter the name of the file to upload, including the directory PATH for the file's location, if needed. The name of the file you're uploading can't duplicate any of those already in the library.

With service-specific software, the process is much less tedious; it's usually reduced to an exercise in pointing and clicking with the mouse (recommended, even if you don't use Windows). In most cases, you're provided with an estimate of the time the transfer will take, and a thermometer-style progress indicator and a constantly updated estimate of the time remaining. The CompuServe Information Manager (CIM) even adjusts its estimate of the time remaining based on the transfer's performance, so far. (The software's initial time estimate is based on file size and modem speed but doesn't make any allowance for your modem's data compression capability, if any.)

For E-mail file transfers, see Chapter 13.

CHAPTER 16

Electronic Bill Paying

Banks have, from time to time, offered online banking services to consumers. When personal computers were still expensive, some even provided a small terminal, with a built-in modem operating at 300 bps.

Today, it's more likely your bank has decided that the Touch-Tone telephone is the best way to support inquiries into account balances and recent debits and credits.

However, with over 2 million households subscribing to PRODIGY, a number of banks have jumped on the bandwagon, and are offering services—including bill paying—to PRODIGY households in their area.

Bill Paying on PRODIGY

The banking institutions listed in Table 16.1 offer regional or local banking services, including bill paying, in the areas shown here.

Table 16.1 Institutions Offering Online Banking Services on PRODIGY

Institution	*Area Served*
Bank One	Columbus, Ohio
Boatmens	Missouri, Tennessee, Kansas, Illinois
Chemical Bank	New York
Chevy Chase	DC area
Comerica	Michigan
Delaware Trust	Delaware
First Interstate	Colorado
First PA Bank	Pennsylvania
Hamilton Bank	Pennsylvania
Mac Banking	Pennsylvania, Delaware, New Jersey
Meridian	Pennsylvania
Midlantic	New Jersey members
NBD	Michigan, Illinois, Ohio
NJNB Homebank	New Jersey members
Provident	Philadelphia
Wells Fargo	California
Wilmington Trust	Delaware

In addition, an organization called BillPay USA offers online bill-paying services to all PRODIGY users.

How Does It Work?

Bill-paying services can make payments in one of two ways—by writing and mailing a check, or by means of *electronic funds transfer* (EFT). In EFT, your account is debited, and the payee's account is credited simultaneously. It's very common in Europe and is finding increasing acceptance in the United States. You may already be paying life insurance premiums and utility bills by direct debit to your account. You may also be paying your bill at the supermarket checkout (and other places) by means of your ATM card.

Single and Recurring Payments

It's more than likely, though, that you spend a fair amount of time each month writing and mailing checks to a number of organizations—from book clubs to department stores and credit card companies. The online bill-paying services are designed to handle this kind of bill paying.

Payments fall into two simple categories—*single* and *recurring*. In either case, all you need to do (once you have signed up for the service) is use the PRODIGY **JUMP** option to get to the service you are using, then enter the payee information, amount, payment date, and payment category. For recurring payments, the date you provide is the starting date; you must also say how often such payments are to be made. As long as you provide the information in sufficient time (for BillPay USA, at least six days before the due date), you are guaranteed that the payment will be made on time. All payments are debited to the bank account number you provide at the time you sign up.

Name That Payee—But Only Once

Once you have provided information on a payee, you can recall it from a list for subsequent payments.

Other Options

Other options include reviewing pending payments, cancelling payments (as long as you do so in good time), and reviewing payment history, either for all payees or for a selected payee.

BillPay USA even has an option to download your payment information, correctly formatted for any of a number of PC financial packages—Managing Your Money, Quicken, Microsoft Money, and MacMoney—or for your favorite spreadsheet program.

CheckFree

Like BillPay USA, the CheckFree payment service (from CheckFree Corporation) is available throughout the United States and can be used with an account at any bank. Unlike BillPay USA, which is available only through PRODIGY, CheckFree doesn't require that you subscribe to an online service. However, if you're a CompuServe subscriber, you can obtain information on the CheckFree service and place an order online for the service and for CheckFree software (just type **go checkfree** and follow the instructions). You can even download a demonstration. If you are not a CompuServe subscriber, you may want to call CheckFree Corporation at 800-882-5280.

To use CheckFree, you need one of several financial management software packages. Apart from CheckFree's own software, the best-known is Quicken, whose sales are enhanced by its CheckFree capability, but don't depend on it. Quicken and the others are sold in all the computer and software stores. The latest versions of Quicken are Quicken 6 for DOS and Quicken 2 for Windows. All the CheckFree-capable packages come with a booklet, describing the CheckFree service, and a sign-up form. We'll use Quicken as our example, but don't take that as an endorsement.

Quicken allows you to print checks, or just maintain a check register (as a basis for writing checks), or it allows you to identify a payment as one that is to be made electronically. Using your modem and Quicken's communications options, you can transmit your payment instructions to CheckFree (before the payment date).

Quicken dials the CheckFree processing center and transmits the bill-paying instructions. The processing center sends back a confirmation number which Quicken enters in its check register.

(If you have a CompuServe account, you may notice that CheckFree uses one of your local CompuServe access phone numbers. Don't worry; they haven't mysteriously signed you up for an additional CompuServe account. The account is theirs.)

CheckFree service is $9.95 per month for up to 20 payments, with the first month free. Additional payments are $3.50 for 10 or fewer (which means that you pay another $3.50 for 1 to 10 extra payments, or $7.00

for 11 to 20 extra payments, and so on). You can only use one bank account with a CheckFree account; however, by special arrangement, you can have up to five CheckFree accounts.

Is It for You?

If you run your finances close to the bone, you're likely to spend much of your time reviewing pending payments and cancelling them. If this is the case, online bill paying may be more trouble than it's worth.

However, if your finances are under control and you usually keep your expenses within your budget, it may be worth trying. Before you sign up, compare the cost of the service with your present cost of writing and mailing checks. You may also want to ensure that the bank used for the payments shows the name of the payee on EFT transactions, and returns all cancelled checks to you.

How Can You Sign Up?

Signing up for any of PRODIGY's bill-paying options is quite simple. Just **JUMP** to **BILL PAYING** to see a list of the banks, plus BillPay USA. Select the one you want to use and follow the instructions as they appear. If you choose one which doesn't provide service in your area, you will receive a message to that effect (and no further information on their bill-paying service). Remember, the banks require you to have an account with them; BillPay USA will use whatever account you designate.

To sign up for CheckFree, either order it through CompuServe (if you want to use CheckFree's own financial software) as described under CheckFree, above, or call CheckFree Corporation's 800 number (also above). If you want to use Quicken or any other CheckFree-capable software (which you can find by browsing your local software store, or computer megastore), just follow the instructions in the CheckFree booklet that will be included.

CHAPTER 17

Remote Computing

Remote computing has meant a number of things over the years. Today, however, it often means using a modem-equipped computer in one location to operate another (compatible) modem-equipped computer in another location. This is accomplished through the use of *remote-control software* loaded on both computers. Such software allows the keyboard, display, and sometimes the printer on the *local computer* (one where the user is sitting) to be used as though they were attached to the other (*remote*) computer.

Remote Computing Software

Various remote-control software packages are available to allow you to work this kind of magic. The three best-known are pcANYWHERE, Carbon Copy Plus, and the aptly-named PC Commute feature of PC Tools.

The LAN at the Office

Why would you want to operate your office computer from afar, rather than just doing the work on your home-based computer and using your modem link to the office to gain access to the files you need?

The answer is two-fold; local area networks and performance. Let's look at local area networks first.

With *local area networks*, or *LANs*, computers are linked together by way of very high-speed connections (as fast as 16 megabits per second) for the purpose of sharing files, printers, and other resources. Typically, the files that are shared reside on disk drives on a *file server*—usually a very fast personal computer (with a great deal of memory and disk capacity) running software such as Novell's NetWare. Each computer on the LAN treats the file server's disk storage as an extension of its own. Some files on the file server are accessible to multiple users (shared files); some are accessible only to their "owner" (private files).

When a LAN-attached computer reads a file or loads a program from the file server, the data is transferred to the computer's memory just as quickly as it would be if it came from the computer's own hard disk drive. At 16 megabits per second, this is hardly surprising. Now let's look at performance.

Suppose a computer's connection to the file server is not at 16 megabits per second? Let's suppose we attach a 14,400-bps modem to the file server. Let's also suppose that, using suitable software, you use your 14,400-bps modem to call up the file server, and attach to it in order to do the same kind of work you did over the 16-megabit-per-second connection at the office. Even if we assume that the modems manage to achieve a fair degree of data compression (using MNP Class 5 or V.42 *bis* capability), every megabyte of information transferred between the file server and your computer's memory would take from three to five minutes. Loading even a modest-sized program would take as long as three minutes.

Suppose we just move the modem from the file server to your office computer (the one you use with the file server). Imagine next that we use

Carbon Copy Plus, pcANYWHERE, or PC Commute (along with your dial-up connection) to attach your home-based keyboard, display, and printer to your office-based computer. Now the information we transfer over the 14,400-bps connection consists only of keystrokes and display information (and if you can't wait until you're back in the office for printed output-information to be printed). When you load a program from the file server, it is transferred to your office computer, and runs there. Similarly, when a program reads from (or writes to) a file on the file server, transfer is only to or from the memory of the computer at the office.

PITFALL!

> The kind of remote-control computing just described may not be suitable for all applications that display information in graphics mode. This is because a large amount of information must be transferred from the remote computer for display on the local computer. The latest versions of the remote-control software, however, use data-compression techniques (and a number of clever tricks) to reduce to a reasonable level the amount of information that must be transferred.

Do You Want to Be a Telecommuter?

We've come a long way from the time when taking work home was just a matter of stuffing some papers in a briefcase. With the kind of connection we've just described, you can spend entire working days at home, saving gas and preserving the environment.

> **WORD POWER**
>
> A *telecommuter* is someone who commutes by telephone, typically using the telephone link to tie a home-based computer into a computer at the workplace. The home-based computer may be used as a terminal, or as a logical extension of a similar computer at the workplace.

Your company might be amenable to such an arrangement. Better still, they might encourage it. Better yet, you might get them to pay for your modem, your remote-control software and even the computer itself. Instead of rushing through your morning ablutions and joining the rat race on the freeway, you can put on your bathrobe and stagger down the hallway from the bedroom to your office just like some of the people who write computer books. Obviously, though, working at home will take some self-discipline.

Is It Really That Simple?

Your part of it is very simple. For the administrator of the LAN at your office, it's not quite as simple as we've depicted. If a number of people are to be telecommuters on a regular basis, there may not be a need for so many computers on the LAN. The LAN administrator will want to consider software products that allow a single personal computer a fast one with lots of memory, and with several attached modems to operate as several "virtual" computers. Such software does exist; installing it and managing will not be your problem unless your interest in computers eventually leads to your becoming the LAN administrator.

If You Have a Laptop Computer

If you have a laptop computer and a home-based computer, both with modems, you can use remote-control products like Carbon Copy Plus, pcANYWHERE, or PC Commute to operate your home-based computer when you're away from home. You just load one copy of the software on the home-based computer before you go, and leave the computer on (or have someone else in the house do it for you when you need it). In your hotel room (or wherever you have access to a phone outlet), you load the other copy of the remote-control software on your laptop, connect its modem to the phone outlet, dial your home-based computer's phone number, and you're in business.

PITFALL!

> Although many hotels now have phones with a "MODEM" socket on the side, too many of them also use "auxiliary" providers of long distance services and, for direct-dial calls, will often add a hefty surcharge to already-inflated rates. To get around this, you will have to follow the manual dialing procedure described in your modem's user guide. That means you might dial not only your own long-distance carrier's access code (for example, 102880 for AT&T) and your home-based computer's phone number, but also your calling card number.

If you want to make use of an online service while away from home, you'll need to do it directly from the laptop. Alternatively, you could install two modems at home—one for the link from the laptop, the other for the link to the online service. Of course, that means another phone line.

CHAPTER 18

Modem Care and Troubleshooting

If you have expensive stereo equipment, you're probably used to taking care of electronic devices. Caring for a modem is as easy as caring for a stereo. You simply need to exercise basic common sense. (Most of the modem care techniques apply to external modems only, because internal modems are protected by the computer's system unit.)

> **COMMON SENSE**
>
> Internal modems require neither more nor less care than any other add-in board for your personal computer. As with other boards, you need to ensure that the backplate securing screw is tight, to keep the unit operating properly and keep dust out of the system unit.

Modems Aren't Coffee-Proof

You probably don't put cups of coffee on your stereo amplifier or audio tape player. There's a good chance, however, that you'll place your modem next to your computer; there's a temptation to treat its top surface like any other convenient surface. It's okay to pile a few diskettes on it (as long as they don't cover any ventilation holes), but your "World's Best Dad/Mom/Brother/Sister/Kid" coffee mug (or any other container of hot or cold liquid) is definitely not welcome.

If you've ever spilled any liquid on your PC keyboard, you'll understand.

Modems Don't Mind Dirt, but They May Mind Dust

Maybe you're fastidious. Maybe you're not. Maybe you're happy to let your keyboard and the bezel of your monitor accumulate 187 layers of dirt. So be it. Your modem can accumulate dirt too; it won't affect the operation. However, it has ventilation holes in the top, dust might accumulate inside. Even though it probably won't affect the operation, you might want to remove the dust (or at least most of it) by blowing into the vent holes or sucking it out with your vacuum cleaner's crevice tool.

PITFALL!

> Whatever you do, don't poke anything through the modem's vent holes; you could damage vital components.

If You *Really* Need to Wipe It Off

Just in case you really are fastidious and don't like your modem to look grubby, wipe it clean with a cloth that is only just damp—*not* dripping with water. If you use a spray cleaner or polish, spray a little on the cloth (*not on the modem*) before wiping.

Modems Don't Like Kids or Animals

Cats love to sit on warm things. When they're switched on, modems get moderately warm. Enough said! If you have a very small child, you may have noticed his or her inclination to pick up everything that can be picked up, sometimes followed by an urge to throw down the newly-acquired object. Modems are small and eminently grabbable. Keep yours out of reach.

Simple Modem Troubleshooting

First you need to know what happens under normal circumstances. Assuming your modem has a speaker and a reasonable set of lights (see Appendix A), you should see and hear the following, starting at the point where you tell the software to dial the call to the online service or remote computer:

1. The modem's DTR (Data Terminal Ready) light comes on, indicating that your software is loaded and has initialized the COM port. The DSR (Data Set Ready) or MR (modem ready) light is also on, indicating that the modem is switched on and ready to be used.

2. The modem's SD (Send Data) and RD (Receive Data) lights flicker dimly. This indicates that the software is sending some initial commands (called a *setup string*) to the modem.

3. The off-hook light, if any, turns on. You can hear the dial tone from the modem's speaker. Your software indicates that it is dialing, and (typically) displays the number being dialed.

4. If you will be connecting (or trying to connect) at 9600 or 14400 bps, the modem's high-speed indicator light (if any) turns on (depending on the approach used by your software, it may turn on in step 9 instead). If your modem setup string (or the modem's default setting) involves forcing CTS (Clear To Send) and, possibly, DCD (Data Carrier Detect) to be on all the time, the CTS and DCD lights turn on. Forcing CTS and DCD on is not unusual for full-duplex communication.

5. You can hear the number being dialed. For a local number, you hear the ringing tone almost immediately. For a different area code, there is usually a short delay. If you hear a busy tone, your software will probably put the modem on hook, and retry (from step 3) with the same or a different number. (Some software, such as Windows Terminal, does not recognize the busy condition; it puts the modem on hook after a certain time—typically, about 45 seconds—and waits for you to restart the operation.)

6. The remote modem answers, and you hear a high-pitched tone. This indicates the start of what is called the *handshaking* procedure between the modems. The tone may, for a fraction of a second, be replaced by one of a lower frequency.

7. The tone is replaced by a rushing sound, indicating that the two modems are *training* (getting in step). If the remote modem is operating at 2400 bps or less, the rushing sound is quite brief. If the remote modem is operating at 4800 or 9600 bps (V.32), or 12,000 or 14,400 bps (V.32 *bis*), the rushing sound lasts two or three seconds, and may change in character after the first second.

8. The modem speaker turns off (although the setup string can include a command to indicate that it is to stay on—and annoy you throughout the communications session). The handshaking procedure is now in its final phase: the negotiation of error-correction and data-compression protocols.

9. If the two modems agree on an error-correction protocol (MNP Class 2, 3, or 4, or CCITT V.42), the error-correction indicator light (if any) turns on. If they agree on a data-compression protocol (MNP Class 5 or higher, or CCITT V.42 bis), the data compression indicator light (if any) turns on. If they were not turned on in Step 4, the CTS (Clear To Send) and DCD (Data Carrier Detect) lights turn on.

10. If your software signs you on automatically, there is now activity on the screen and at the modem. On the screen, you will see progress messages. At the modem, you will see a great deal of flickering of the SD (Send Data) and RD (Receive Data) lights. If you have to sign on, you may see the RD light flicker, and a message appear on your screen. As you sign on, you will see both

the SD and RD lights flicker as you enter text, and as each character entered is echoed by the remote system. Whenever the remote system sends you a response, you will see the RD light flicker.

11. As you interact with the remote system, you will continue to see SD and RD light activity. If you are using terminal emulation, the RD and SD lights will both flicker when you are using the keyboard; only the RD light will flicker when the remote system responds. If you are using special software, such as CIM, OzCIS, TAPCIS, or the PRODIGY or America Online packages, the SD light will flicker when you select a menu item (or otherwise tell the software to act on your behalf). The RD light will then flicker, indicating that the response is arriving from the remote system.

12. Whenever you upload or download files, either the SD or RD light will flicker. For uploads, the SD light will flicker furiously, while the RD light flashes intermittently. For downloads, the roles are reversed. The intermittent flashing corresponds to acknowledgments that each data packet has either been received intact, or needs to be retransmitted. Depending on the file-transfer protocol and the use of data compression, the CTS (Clear To Send) light may occasionally turn off, then back on during an upload. (This would happen, for example, if you specified a COM port speed of 38,400 bps and modem operation at 9600 bps. Whenever the modem cannot achieve a 4-to-1 compression ratio, it must make your computer wait in order to avoid losing data.)

13. Finally, when you sign off, the modem's off-hook light turns off, along with most of the other lights (except DSR and DTR). When you exit (or unload) your communications software, the DTR light turns off.

Possible Sources of Trouble

In spite of all the places where something could fail, problems are rare. But, when they do occur, the source of trouble might be any of the following.

- Your computer.
- Your communications software.

- Your modem cable.
- Your modem.
- The cable from your modem to the wall jack.
- The wall jack.
- The wiring from the wall jack to the phone company's standard network interface.
- The telephone line to the phone company's local office.
- The telephone circuit between your end and the online service or other remote computer.
- The online service (either equipment or software).

Possible Timing of Trouble

Problems can appear at any of the following times:

- During call setup (between the start of dialing and the establishment of a connection).
- Intermittently during a communications session.
- During all communications sessions.
- During communications sessions with a particular online service or remote computer.
- During file transfer (with a particular protocol or with all protocols).
- At the conclusion of a communications session (for example, failure to disconnect).

Information Sources

There are several good sources of troubleshooting information. They include:

- Your common sense.
- Your communication software's error messages.
- Your communication software manual.
- Your modem's indicator lights.
- Your modem manual.
- The online service.
- A friend who knows more than you.

We'll deal with the common sense. Your manuals and your software's help functions will handle the rest.

Using Your Head

Here are some obvious modem problems you can diagnose for yourself.

Symptom: External modem is powered on, but does not respond to your software.

Likely problem: Insecure (more likely) or defective (less likely) or incorrect (less likely) modem cable, or defective modem (possible).

Solution: If the modem has responded before, make sure the cable is securely connected to the modem and to your COM port, and that there is no looseness where the cable wires enter each of the connectors. You may even want to try a friend's modem cable in place of your own. If the problem persists, turn to the troubleshooting section of your modem manual. You may have to run modem diagnostics (not as hard as it sounds), which will reveal any modem defects.

If you are going through your initial hardware and software installation, check to make sure you have the right modem cable (if necessary, where you bought it). Also check to see that you have configured your software properly (correct COM port?).

Symptom: The modem (internal or external) responds, but when it goes off hook, there is no dial tone from the speaker. (Even if the modem has no speaker, your software will probably also indicate that there is no dial tone.)

Likely problem: Wiring. (Or you didn't pay your phone bill.)

Solution: Check the modular cord: does it indeed connect the modem to the wall jack? If it does, unplug it from the modem, and plug it into a telephone. If the phone works, you have a modem problem. If the phone doesn't work, try the phone and cord with a different wall jack. If the phone works now, your problem is in the wiring to the wall jack you were using for the modem. If the phone still doesn't work, try another modular cord (from a phone that is working, for example). If the phone works now, trash the original modular cord and get a new one.

If the jack is one of two or more serving the same phone number (all on line 2, for example), and a phone doesn't work when plugged into any of them, the problem is either in the wire from the first jack to the phone company's standard network interface, or with the phone company. (See Chapter 6 for more details on phone-line wiring.) Open the door of the standard network interface box, remove the modular plug on the left side (if line 1 is the problem), or the right side (if line 2 is the problem). Plug the phone into the socket from which you removed the plug. If the phone works, the problem is in your wiring. If it doesn't, call the phone company.

Symptom: When you are using the keyboard in terminal-emulation mode, what appears on the screen is garbage. Also, more characters may appear on the screen than you could possibly have entered.

Likely problem: Poor connection. The extra characters are the remote computer trying to tell you that your input is meaningless.

Solution: If it only happens occasionally, hang up (or tell your software to hang up) and place the call again. If it happens frequently, it may be a wiring problem. Make some normal phone calls on the same line to see if the line is noisy. If it is, plug the phone into the socket in the standard network interface box. If it's still noisy, call the phone company. If not, check your internal wiring. If it is the phone company's problem, they may want to change your number to one with a prefix that connects you to a more up-to-date (and more reliable) switching office.

> **COMMON SENSE**
>
> If you add a second line to your modem, ask the phone company to give you a number served by their newest central office in your area. Tell them why. In metropolitan areas, a single street address may have access to several prefixes.

Symptom: File transfers keep failing before completion. There may be an error message (like **File Transfer Aborted**), or the process may just "hang" (do nothing).

Likely problem: Mismatched protocols, incorrect flow control, or deteriorating connection.

Solution: For mismatched protocols, the file transfer will fail almost immediately. For general-purpose software check that the protocol you have specified (XMODEM, XMODEM-CRC, Kermit, etc.) is exactly the same as the one being used at the other end.

If the file transfer fails part way through, the connection may have deteriorated or you may not have configured your software for flow control. Your software manual will tell you about *X-On/X-Off* and *RTS/CTS flow control*, and when to choose one or the other.

If you're using automated software such as CIM, OzCIS, TAPCIS, or the PRODIGY or America Online packages, file-transfer failure is unlikely (the protocols always match)—except where the connection is poor. If it occurs frequently, or always at the same point in the transfer, contact the online service. For CIM, OzCIS, or TAPCIS, you may want to post a message to the appropriate CompuServe forum (DOSCIM, IBMCOMM and TAPCIS, respectively), describing the symptoms. Address your message to SYSOP.

Note that file transfer is more sensitive to a poor connection than is terminal emulation. This is because you are more likely to encounter noise during the transfer of a packet of 128 bytes or more than you are during the transfer of a single character. (Your keystrokes are transmitted as you type them.)

Now It's Your Turn

The examples above are intended to help you develop a logical approach to problem solving. Other problems might be the result of incorrectly specified AT command strings, or software otherwise configured incorrectly.

This is the final chapter of the book. Only the appendixes remain. Our last piece of advice to you is: Read the Manual. If you acquaint yourself with the troubleshooting sections of your modem manual, your software manual, your online service guide, and so on, you'll be well equipped to recognize symptoms and take the right corrective action. (Knowing all the toll-free help line numbers doesn't hurt, either.)

May all your troubles be little ones.

APPENDIX A

Modem Indicator Lights and What They Mean

Most modems are equipped with a set of indicator lights. Their purpose is to show the status of the modem, and of the various interchange circuits which the modem and your computer use to signal one another or to transfer data.

The way the lights are labeled varies from manufacturer to manufacturer. The table below uses commonly accepted abbreviations for the labels, with some common variations. Not all modems have all the lights listed, and some modems have lights that are not listed here.

Label	*Interchange Circuit*	*Direction*	*Explanation*
AA	No		Auto answer enabled.
CTS	Yes	From modem	Clear To Send. Indicates to the PC that it may transmit over the Send Data circuit (SD). If on constantly, indicates full-duplex operation. Sometimes abbreviated CS.
DC	No		Data compression enabled. May be MNP Class 5 (or higher), or V.42 *bis*.
DCD	Yes	From modem	Data Carrier Detect. Flashes rapidly when the modem is communicating with the remote modem (even when no data is being sent or received). Sometimes abbreviated CD (Carrier Detect).
DSR	Yes	From modem	Data Set Ready. Indicates that the modem is powered on and ready. Sometimes abbreviated MR (Modem Ready).
DTR	Yes	From COM port	Data Terminal Ready. Indicates that the PC is powered on and ready. Sometimes abbreviated TR (Terminal Ready).

Label	*Interchange Circuit*	*Direction*	*Explanation*
EC	No		Error Correction enabled. May be MNP Class 2, 3, 4, or higher, or CCITT V.42.
HS	No		High-Speed operation. Indicates that the modem is operating at one of the higher speeds—typically, one of the speeds associated with V.32 or V.32 *bis* operation.
OH	No		Off Hook. Remains on as long as the modem is initiating a call, answering a call, or communicating.
RD	Yes	From modem	Receive Data. Flashes rapidly when data is being received. Sometimes shown as RX.
RI	Yes	From modem	Ring Indicator. Indicates that the modem is requesting the PC to answer an incoming call.

continues

Label	Interchange Circuit	Direction	Exp
RTS	Yes	From COM port	Request To Send. Indicates that the PC has requested clearance to transmit over the Send Data circuit (SD). May be on constantly for full-duplex operation. Sometimes abbreviated RS.
SD	Yes	From COM port	Send Data. Flashes rapidly when data is being transmitted. Sometimes shown as TX (Transmit) or TD (Transmit Data).

Your modem manual will explain the abbreviations used by its manufacturer, and should provide a full explanation of each light's function. Commands your software has sent to the modem (or the modem's default settings) may force some lights to be on constantly (Data Carrier Detect and Clear to Send, for instance).

APPENDIX B

The Hayes AT Command Set

Hayes Microcomputer Products has defined a large number of modem control commands—now known as the *Hayes AT Command Set.* Your communications software uses them, and (if you are using a terminal emulator) you can also use them to control your modem directly. Hayes-compatible modems may use a few or very many of them—we refer to anything less than the whole command set as a *subset.* Although you might expect to pay more for a modem that implements a large subset (many commands) than one that implements a smaller subset (fewer commands), this isn't always the case.

This appendix describes some of the commands implemented by the Eagle+ modem, from Digicom Systems, Inc. We chose the Eagle+ not to endorse it, but because of the large number that were sold (to members of the Prodigy Service) as a result of an advertising blitz on PRODIGY.

- All the AT commands supported by your modem will be listed and described in its manual.
- You should note that modems are not *case-sensitive* when they are in command mode; this means you can type your commands (or specify them when configuring your software) using either uppercase or lowercase letters.
- To understand how the commands work, examine the examples carefully, noticing how they are put together.

The Initial AT

The AT commands are made up of letters, numbers, and other characters (such as plus signs and ampersands). "A" and "T" are called the *attention characters*—your modem's "wake-up call." Use them to precede all other commands except one (the second in the list that follows). AT may precede a single command or a series of commands (called a *command line*), but you must always end your command with a carriage-return character (by pressing **Enter**). In your software's configuration menus, normally you will find the characters **^M** to signify a carriage return.

The Commands

A

This one means, "Answer incoming call immediately." Suppose your modem is in auto-answer mode, and your software is awaiting a call (in standby mode). If you've turned on the Ring Indicate circuit at the interface, the software will inform you when the phone is ringing; then it will answer the call by sending this command to the modem in the string **ATA**. If you are using a terminal emulator and awaiting a call (from a friend who is also using a terminal emulator, for example), you can respond to the appearance of the word RING on your screen (or to the sound of ringing, if you have a phone on the same line) by typing **ATA**.

A/

This command is the exception that need not (and *must not*) be preceded by AT. It causes the modem to execute the last command line one more time.

D

D is the dial command. Tone dialing is the default setting. Other commands can be used after D to modify the way it works:

T and **P** specify tone and pulse dialing, respectively. The sequence **ATDP 555-1234** will use pulse dialing (at the rate of 10 pulses per second) to call the number 555-1234. The space, the hyphen, and (where used) any parentheses (as for area code) are ignored.

W limits the wait for a dial tone. The length of the wait is determined by the value in one of the modem's *S registers* (explained later in this appendix), specifically S7. If the value in S7 is 45 (the default), the sequence **ATDTW 555-1234** would cause the modem to go off hook, wait up to 45 seconds, and (if a dial tone at least 3 seconds in duration occurs) use tone dialing to call the number. After 45 seconds, the modem would go on hook and send the response NOáDIALTONE to the computer. Most communications software uses this technique.

A **comma** (,) is used to pause between dialed digits. It's useful for calls made through a PBX system (where you must first dial 9), or after an initial sequence such as *70 (which turns off call waiting). The value in Register S8 determines the length of the pause, in seconds. The sequence **AT S8=1 DT *70,555-1234** would set the pause duration to 1 second, dial *70, pause (for 1 second), then dial 555-1234. The sequence **ATDT 9,,P 555-1234** would dial 9, using tone dialing, pause for twice the time currently specified in Register S8, then pulse-dial 555-1234.

An **exclamation point** (!) is used to cause the modem to simulate a "hook flash," by going on-hook for half a second, then off-hook again. It allows call transfer (and similar functions) under software control.

A **semicolon** (;) returns the modem to command mode after dialing. (It is normally in data mode immediately after dialing.) This is used by software that dials voice calls (Windows Cardfile, for example). When Windows Cardfile is asked to autodial a number (555-1234, for example), it tells you to pick up the phone, pauses about one second, then sends the modem the sequence **ATDT 555-1234;**. It then displays an OK button. When you click on **OK** (after dialing is complete), Cardfile sends the sequence **ATH** to the modem to put it on-hook. Because your phone is now off-hook, you won't lose the call.

@ (the "at" character) causes the modem to "wait for quiet," specifically 5 seconds of silence. This is useful for secure computer systems where they insist on calling you, after verifying that your phone number is on their authorized list. Let's say the computer center's number is 555-1234, and your number is 555-6789. The sequence **ATDT 555-1234 @ 555-6789 ,;H** would dial 555-1234, wait for five seconds of silence after detecting one ring, dial 555-6789, pause (normally for 2 seconds), switch to command mode, and go on-hook. Meanwhile, the computer at the center would have recorded the number 555-6789, checked it against a list of authorized numbers, then (after a delay of a few seconds) would place a call to the number (also by way of a Hayes-compatible modem). Your software would be waiting for the RING message from that modem, at which time it would complete the connection by sending the modem the sequence **ATA**.

S is used to dial one of 10 stored numbers, referenced as 0, 1, 2 ... to 9. The sequence **ATDS3** would, for example, dial the number stored in position 3. The number is stored, in the first place, with the &Z command (described later). **AT&Z3 555-1234** would store 555-1234 in position 3. Each position can accommodate a number of up to 36 digits. Stored numbers are not lost when the modem is turned off.

+++ (Default Escape Sequence)

An *escape sequence*, consisting of three consecutive plus signs, will switch the modem from data mode to command mode. It is always preceded by at least one second of "dead time" (nothing being sent to the modem), so that the occurrence of +++ in data will not cause the modem to switch modes. It is the ASCII value stored in register S2 that determines which character will be used for the escape sequence. The value, 43, corresponding to the plus sign, is the default. There is normally no reason to change the escape sequence; if you wanted to change it, however—to three pound signs (###), for example—you would use the command **ATS2=35** (35 being the ASCII value for the pound sign). If register S2 contains a value of 128 or more, the escape sequence capability is disabled.

E0 and E1

E0 disables command echoing; **E1** enables it. *Command echoing* allows you to see your commands on-screen as you enter them. It's useful if you are going to enter commands manually (using a terminal emulator, for example).

H0 and H1

H0 causes the modem to disconnect and go on hook (that is, to hang up); **H1** forces the modem off hook (though this is not required for dialing). H0 may be shortened to **H**.

I0 to I8

Each of these commands will cause the modem to respond with certain information. **I7** is the most interesting; it produces a summary of all the AT commands supported by the modem.

K0 and K1

As a user, you could interrupt the transmission or processing of data with a *break signal*. When your modem is in error-correcting mode, **K0** disables the detection of a break signal; **K1** enables it. See your modem's manual for more details.

M0 to M2

M0 turns the modem's speaker off; **M1** turns it on for dialing, and off again when the handshake with the remote modem is complete; **M2** turns it on all the time. (M2 is not recommended; it will drive you crazy.)

Q0 to Q2

Q0 turns on your modem's responses (OK, CONNECT, and so on); **Q1** turns them off; **Q2** makes the modem provide responses only when it is originating calls.

V0 and V1

V0 makes your modem give short (numeric) responses; **V1** makes it give responses in the form of words.

X0 to X5

These commands indicate whether or not dialing is to be *blind* (no dial-tone detection), whether or not a "busy signal" is to be detectable, and (for each possible combination) whether or not the CONNECT message is to indicate the speed (for example, CONNECT 9600).

&B0 to &B2, &C0 to &C2, &D0 to &D3, and &R0 to &R2

These commands all relate to the control of the signals at the interface between the modem and your computer.

&F

&F resets the modem to the way it was configured when it left the factory.

&S

&S causes the modem to send your computer a list of all stored telephone numbers.

&T0 to &T8

These commands are used to test the modem. If you run into problems, you might want to acquaint yourself with these.

&W

If **&W** is placed at the end of a command string, any options changed by commands in the string will be stored. That way they will be saved even if you turn the modem off and on again.

&Z0 to &Z9

These commands are used to store phone numbers in positions 0 to 9. For example, **AT&Z4555-1234** will store the number 555-1234 in position 4.

*E0 to *E10

These commands are used to specify options for error control and data compression.

*F0 to *F3

These commands specify how flow control is to be implemented.

*N0 to *N6

These commands specify the initial connection's *data rate* (speed of transmission, in bits per second). *N0 to *N6 specify, respectively, 1,200, 2,400,

4,800, 9,600, 7,200, 12,000, and 14,400 bps. The actual rate may depend on the capabilities of the remote modem.

*O

*O causes the modem to display (send to the computer) its current *operating parameters* (all the values that can be set with any of the commands).

*S0 and *S1

***S** or ***S0** turns off speed conversion; ***S1** turns it on. *Speed conversion* allows your computer to send data to (and receive data from) the modem at a rate greater than the modem's transmission speed. This allows you to take advantage of data compression.

S Registers

Hayes-compatible modems have *registers* (special memory locations), each capable of holding a one-byte value (8 bits) which can range from 0 to 255. They are referred to as *S registers* (S0, S1, S2, and so on). A partial list, including a description of each register's purpose, follows.

S0 specifies the number of times the phone can ring before the modem answers (when in auto-answer mode). If the value in S0 is zero, auto answering is disabled.

S1 is used in auto-answer mode, to record the number of rings that have occurred.

S2 contains the ASCII value that the modem will recognize as the *escape character* (which returns the modem to command mode from data mode). The default is the ASCII value 43, the plus sign (+).

S3 contains the ASCII value that the modem will recognize and use as a *carriage-return character*. The default value is 13 (the actual ASCII value for a carriage return, corresponding to your computer's Enter key).

S4 contains the ASCII value that the modem will recognize as a *line-feed character*. The default value is 10 (the actual ASCII value for a line feed). When the modem sends responses in the form of words, they are followed by a carriage return and a line feed.

S5 contains the ASCII value that the modem will recognize as a *backspace character*. The default value is 8 (the actual ASCII value for a backspace). The modem uses backspace characters to erase preceding characters (in a command sequence) from right to left.

S6 contains a value equal to the maximum time, in seconds, the modem is to wait before dialing.

S7 contains a value equal to the maximum time, in seconds, the modem is to wait for completion of the handshake with the remote modem. This value is also the maximum time the modem is to wait for a dial tone when there is a W command in the dial sequence.

S8 specifies the length, in seconds, of the pause to be taken when a comma is encountered within a phone number.

S*r*? Commands

S, followed by the number of an S register, followed by a question mark, will cause the modem to display the current value in the specified register. To use the command, you type the register number instead of the ***r***. If, for example, you wanted to know whether your modem was set for automatic answering, you would type **ATS0?**. The modem would respond with the value (0, for no auto-answer, 1 or greater for the number of rings if auto-answer mode is set) in register S0, followed by a carriage return, followed by **OK**.

APPENDIX C

Glossary

This glossary deliberately includes more terms than are used in this book, and certainly more than you probably need to know. You will, however, encounter many of them as you read through your modem and software manuals; you don't want them to remain a total mystery to you. No offense is intended to the writers of such manuals, but the definitions they usually provide are often a little brief and somewhat uninformative, and sometimes even a little bit inaccurate. (Note: Many of the definitions include the generic terms DTE and DCE. Look these two up first.)

analog signal The signal one modem uses to communicate with another modem.

ANSI American National Standards Institute. Standards-setting organization and the official U.S. member of ISO (International Organization for Standardization).

ASCII American Standard Code for Information Interchange. Code in which each of 128 possible characters is defined by a combination of seven bits. Normally, of the 128 values, 32 are reserved for control characters, 94 for printable (displayable) characters, one to represent a space, and one for an idle character. Personal computers use an IBM-defined extension of ASCII, which uses eight bits for each character and has 256 possible values. Some communication services use only the basic, seven-bit version of ASCII.

asynchronous transmission See *start-stop transmission.*

auto-answer Modem feature which, when activated, allows the modem to alert the computer to an incoming call, at which time the computer (if running appropriately configured software) will complete the connection. All modems use the Ring Indicate circuit (RI) for this purpose. Hayes-compatible modems also use the Receive Data circuit (RD) to send the word RING (or, if set up for short messages, the digit 2) to the computer.

baud Unit of measure used in specifying the speed of transmission between two modems. One baud is equal to one change of state per second. "States" (or changes of state) are used to represent individual bits or groups of two, three, four, or more bits. Baud and bits per second are not the same thing! Note that when manufacturers and resellers of modems and related communications products say "baud," they really mean "bits per second."

baud rate Informal (but incorrect) term for *bit rate*. See also *baud.*

BBS Bulletin board system. A small online service, usually operated by one person, a couple, a special interest group, or a manufacturer (for product support). See Chapter 14.

Bell 103 Modem standard, developed by the Bell Telephone Company, for full-duplex start-stop transmission at up to 300 bps.

Bell 212A Modem standard, developed by the Bell Telephone Company, for full-duplex start-stop transmission at up to 1200 bps. Most of today's modems, designed for operation at 1200 bps or faster, support this standard.

binary transfer See *file transfer.*

bit The smallest possible data element, capable of representing either of two possible values—zero and one. (The term is an abbreviation of "binary digit.")

bit rate Transmission speed, expressed in bits per second.

bit/s Standard abbreviation for bits per second.

bits per second Unit of measure for data transmission. Correctly abbreviated to bit/s (for example, 2400 bit/s). Often informally written as bps, BPS, or b/s. This book uses bps by popular demand.

bps Popular abbreviation for bits per second.

byte Data element consisting of eight bits and having any of 256 possible values. May represent a printable or displayable character (for example, letter, numeral, punctuation, or other symbol), a numeric value, or a part of a larger numeric value (that is, one exceeding 8 bits in length), or anything else.

carrier signal Continuous signal from one modem to another (over the telephone circuit), capable of being modulated whenever there is data to transmit. Also called "modem carrier."

CAS Communications Application Specification. An Intel standard defining how a computer and a fax modem are to coordinate the sending and receiving of facsimile information to and from Group 3 fax machines. CAS is a popular de facto standard, but not an accepted industry standard.

CCITT Consultative Committee, International Telegraph and Telephone. Organization, within the International Telecommunications Union (ITU), which recommends standards for data communications. CCITT Recommendations are numbered in several series, each of which has its own designated initial letter. Many of the V-Series Recommendations (for example, V.32) deal with modems. (The correct expansion of CCITT is, in fact, French—Comité Consultatif International Télégraphique et Téléphonique.)

Class 1 fax modem A modem that conforms to the EIA-578 standard, which defines how a computer and a fax modem are to coordinate the sending and receiving of facsimile information to and from Group 3 fax machines.

Class 2 fax modem A modem conforming to EIA-592 standard, which serves the same purpose as EIA-578, but redefines the division of work between the computer and the modem (the modem does more than with Class 1).

CTS Clear To Send. Indication from the DCE to the DTE that data may be transmitted. Used in half-duplex operation. See also *RTS*.

COM port As viewed from the back of a PC, one of the connectors (typically having 9 pins, but sometimes 25 pins) used for connection to a serial communication device, such as a modem or a printer (although most printers use a parallel interface and connect to an "LPT" port). PC software addresses COM ports as COM1 or COM2 (or, if you've added them, COM3 or COM4). If you have an internal modem, your software will address it as a COM port (typically COM3); however, instead of a 9-pin or 25-pin connector at the back of the PC, there will be an RJ-11 modular socket for the cable to the telephone wall socket, plus a second one for the cable to your telephone set.

data bits The number of bits making up a single character. This number depends on the type of equipment. For most computers, including personal computers, there are eight bits per character; we refer to 8-bit characters as bytes. For ASCII terminals, there are 7 bits per character. The old Telex terminals had 5 bits per character (very few of these are still in use) .

DB9 connector Connector with a row of four and a row of five pins (DB9-P) or sockets (DB9-S) enclosed by a metal shell in the shape of an elongated D. A PC's COM port usually has a DB9-P connector.

DB25 connector Connector with a row of twelve and a row of thirteen pins (DB25-P) or sockets (DB25-S); these are enclosed by a metal shell in the shape of an elongated D. A PC's COM port may have a DB25-P connector; the serial communications port of a terminal almost always has a DB25-P connector. The connector on an external modem is a DB25-S.

DCD Data Carrier Detect. Signal from the DCE to the DTE, indicating it is receiving a carrier signal from the DCE at the other end of the telephone circuit.

DCE Data Circuit-terminating Equipment. A modem or similar device. (The only DCEs this book deals with are modems.)

digital signal The signal a personal computer's COM port (or any DTE) sends to or receives from the modem.

download Term used customarily to describe the transfer of a file from a larger system to a smaller system (e.g., from an online service to a personal computer). May be used as a noun or a verb.

DSR Data Set Ready. Signal from the DCE to the DTE, indicating that it is powered on and ready for data communication.

DTE Data Terminal Equipment. Device at which transmission originates or terminates. When it is using its COM port, a personal computer is a DTE. When it is using an internal modem, it is both a DTE and a DCE (see *DCE*).

DTE-DCE interface The interface between a DTE (e.g., a personal computer) and a DCE (e.g., a modem). Includes transmitted data (SD) and received data (RD) circuits and a number of control circuits—Data Terminal Ready (DTR), Data Set Ready (DSR), Request To Send (RTS), Clear To Send (CTS), Data Carrier Detect (DCD), and Calling (or Ring) Indicator (RI). See *CTS, DCD, DSR, DTR, RD, RI, RTS,* and *SD.*

DTR Data Terminal Ready. Signal from the DTE to the DCE, indicating that it is powered on and ready for data communication.

Duplex See *full-duplex.*

EAASY Sabre Consumer-oriented version of American Airlines' Sabre reservation system, accessible as an option on most online services (America Online, CompuServe, Delphi, GEnie, Prodigy, etc.). Users must have an AAdvantage (frequent flyer) number, but can apply for it online.

EIA Electronic Industries Association. U.S. standards-setting organization. Several EIA standards are concerned with data communications. See *EIA-232-D,* below.

EIA-232-D The best-known data communications standard. It defines the interface between a DTE (for example, a PC) and a DCE (for example, a modem). Specifically, it defines some circuits (send, receive, etc.), and the nature of the electrical signals on those circuits. It is often referred to as RS-232 or RS-232-C; however, the EIA has not used the prefix "RS" since January 1986, and EIA-232-C was replaced by EIA-232-D in January 1987. All external modems offered for use with personal computers conform to this standard, as do a PC's COM ports. If a PC has an internal modem, this standard is not relevant.

E-mail Message or messages sent from one user to one or more other users on a computer network.

external modem A modem that does not reside in a PC (or any other computing device). It must be connected to a DTE (e.g., one of a PC's COM ports) by means of a modem cable.

fallback rate A modem speed that is lower than its normal (that is, maximum) speed of operation. May be used when communicating with a slower, compatible modem, or to facilitate transmission over a line that is too noisy for full-speed operation.

female connector Connector that uses sockets. The connector on an external modem is female, as is the one at the COM port end of the modem cable. Socket positions are customarily referred to by "pin number," rather than "socket number."

file transfer The copying of files from a device at one end of a communications link to a device at the other end of the link. (See Chapter 15.) File transfers may be binary (data content doesn't matter) or text (ASCII characters only).

forum A section within an online service (such as CompuServe, America Online, etc.) where messages may be exchanged under various topic headings within an area of common interest. Forums may also include a library section, from which various files (programs, bug fixes, text, and so on) may be downloaded. Many forums also include one or more "conference rooms" which users may "enter" for online conversations.

full-duplex Mode of operation in which data may be transmitted and received simultaneously. Modems for use with personal computers are almost universally full-duplex. The personal computer software used with them may operate in either half-duplex or full-duplex mode. (If this is confusing, think of a two-way street. Traffic is permitted, but not compelled, to travel in both directions at the same time.)

Group 1 Obsolete international facsimile communication standard, supporting transmission at the rate of one page every six minutes. Very rarely supported by today's fax machines.

Group 2 International facsimile communication standard, supporting transmission at the rate of one page every three minutes. Still supported by most Group 3 fax machines.

Group 3 Current international facsimile communication standard, supporting transmission at rates faster than one page per minute (achieved by using data compression). All fax machines sold today are classed as Group 3. One aspect of Group 3 is the transmission of the rasterized document image at a rate of 9600 bps, with fallback rates of 7200, 4800, and 2400 bps. All data/fax modems support Group 3 operation, but not all of them support the higher speeds. Manuals for such modems may also contain references to V.27 *ter*, V.29, and EIA-578 (plus, possibly, EIA-465 and EIA-466). See also *rasterizing*.

Group 4 New international facsimile communication standard, supporting transmission at rates up to 19,200 bps over voice-grade lines, and up to 64 kbps over digital communication circuits.

half-duplex Mode of operation in which data may be transmitted or received, but not both at the same time.

handshake All the procedures that take place from the moment a connection is established between two modems to the moment at which data transmission is possible. Includes synchronization and *link negotiation*.

Huffman encoding One of the techniques used for data compression. MNP Class 5 supports Huffman encoding and another technique called "run-length encoding."

interchange circuit Any of the circuits between a DTE and a DCE. See *DTE-DCE interface.*

internal modem A modem on a card (printed-circuit board) that occupies one of the expansion slots in a PC. It may be connected directly to a telephone outlet by means of a standard telephone modular cable.

kbit/s Standard abbreviation for kilobits per second. A kilobit is 1000 bits. Often used in specifying the speed of faster modems (for example, 14.4 kbit/s, rather than 14,400 bit/s). A kilobit is 1000 bits. [Note that, for computer memory (RAM), the prefix "kilo" (k) is used to indicate 1024. Thus, in RAM, 1 kilobyte (kbyte) is 1024 bytes.]

kbps Popular abbreviation for kilobits per second. Note that the use of a capital K (Kbps) is incorrect.

LINE Label on one of the RJ-11 sockets of a modem. The LINE socket is connected by modular cord to a telephone wall jack. See Chapter 8.

link negotiation Process by which a pair of modems agrees on transmission speed, modulation method, and protocols to be used between them. Part of the modem *handshake* process.

loopback test See *V.54.*

lurker A person who "hangs around" online bulletin boards and forums, browsing through the messages and, if so moved, replying to some of them. Lurking is a perfectly respectable practice, and the term is not used in any derogatory sense.

lurking See *lurker.*

male connector Connector that uses pins. COM port connectors are male, as is the connector at the modem end of a modem cable.

MNP Microcom Networking Protocol. A set of error-correcting, data-compression and (sometimes) link-negotiation techniques for modems. MNP Classes 1, 2, 3, and 4 deal with error correction. MNP Class 5 specifies two types of data compression. Class 4 includes Class 3, and Class 5 includes Class 4 (therefore, also Class 3). There are higher-numbered classes, but they (and Class 1) are rare in modems sold for use with personal computers.

modem Modulator/demodulator. A device which converts digital signals to a form that can be transmitted over a telephone line (modulation) and which converts the received signals from a similar device to digital form (demodulation). A modem's capabilities may include dialing outgoing calls (and answering incoming calls) automatically, compressing and decompressing data, and correcting transmission errors. Some modems can also handle the transmission and reception of faxes.

modem cable Cable used to connect a modem to the COM port of a personal computer (or other DTE). The end of the cable which plugs into the modem has a 25-pin male connector. The end that plugs into the COM port will typically have a 9-pin female connector, but may also have an alternative 25-pin female connector, or be supplied with a 9-pin to 25-pin adapter.

modular cable (modular cord) Standard cable used to connect a telephone, an answering machine, a modem, or a fax machine to a telephone outlet (which is typically a wall-mounted socket). Manufacturers of telephones and modems supply this cable. Replacement cables in a variety of lengths may be obtained from any store which sells telephones or modems.

modular connector The plug at the end of a telephone cord, or the socket into which it is inserted. In the U.S. and Canada, RJ-22 connectors are used on the cord between the handset and the base unit, and RJ-11 connectors are used on the cord between the base unit and the wall socket. The cable connecting a modem to the wall socket also uses RJ-11 connectors.

modular cord See *modular cable.*

negotiation See *link negotiation.*

network Two or more locations connected for the purpose of exchanging data. Networks such as CompuServe have many interconnected locations.

parity In start-stop transmission, an extra bit that may be added to any character that has seven or fewer data bits. Odd and even parity provide for very elementary error detection (by the DTE, not by the modem). For odd parity, the extra bit will be a one if the total number of "one" bits in the character is even, and will be a zero if the total number of "one" bits is odd. Thus, the total number of "one" bits, including the parity bit, is always odd. If an arriving character, including the parity bit, has an even number of "one" bits (as a result of noise on the line, or some other "glitch"), it is considered to be in error. For even parity, the odd/even rule is simply the reverse of the rule for odd parity. "Mark parity" and "space parity" don't provide any error checking. They are just used to increase the number of bits by one (for example, to pad a 7-bit character out to eight bits). A "one" bit is used for mark parity and a "zero" bit for space parity.

PHONE Label on one of the RJ-11 sockets of a modem. The modular cord attached to a telephone set may be plugged into the PHONE socket, allowing the modem and the telephone to share the same wall jack. There is no requirement that a phone be plugged into this socket. For a two-line phone (and, obviously, a two-line wall jack), the operation of the modem will not affect the use of line 2. See also Chapter 8.

post To place a message on a computer bulletin board or the message section of an online forum. Also used, as a noun, to describe the posted message itself. (Not to be confused with the U.K. English usage of the term, which is synonymous with mail.)

protocol An agreement between parties on the format, meaning, and sequence of control messages to be exchanged between them. The "parties" may be a pair of DCEs, DTEs, or, for file transfer, programs. There are protocols for error correction (MNP and V.42), for data compression (MNP Class 5 and V.42 *bis*), and for file transfer (XMODEM, YMODEM, ZMODEM, Kermit, etc.). (Yes, there are also protocols for controlling data links, but they are not supported by a computer's COM ports, so this book doesn't deal with them.)

rasterizing The process by which a document image is converted to a stream of bits representing either black or white, or one of sixteen levels of gray, for each element of the image. For Group 3 faxes, there are either 98 or 196 raster lines per vertical inch, with a horizontal resolution of 203 lines per inch (yielding 1.86 or 3.72 million elements per 8.5-by-11-inch page). Sixteen levels of grey ("halftone" setting—requiring four bits per element, rather than the one bit required for black and white) can be specified, but are not typically used for documents containing only text and/or line drawings. The bit stream is compressed for transmission, and decompressed when received.

RAM Random-access memory. The internal memory of a computer, within which programs are executed and information is manipulated.

RD Receive Data. The interchange circuit used for the transfer of data from the DCE to the DTE. The abbreviation RX is sometimes used instead of RD.

RI Ring Indicator. Signal from the DCE to the DTE, indicating the presence on the telephone line of a ringing signal (that is, the signal that would normally cause the telephone to ring).

RTS Request to send. Signal from the DTE to the DCE, asking for clearance to transmit data. The DTE will not proceed until it senses a CTS (Clear To Send) signal from the DCE. See also *CTS.*

RJ-11 See *modular connector.*

RJ-22 See *modular connector.*

RS-232 See *EIA-232-D.*

RS-232 cable See *modem cable.*

SD Send Data. The interchange circuit used for the transfer of data from the DTE to the DCE. The abbreviations TD (Transmit Data) and TX are sometimes used instead of SD.

signal ground Common return connection for all other circuits at the interface between a DTE and a DCE.

start-stop transmission Transmission mode in which the start of every transmitted character is signified by a "start bit," and the time between characters is filled with one or more "stop bits." Also called asynchronous transmission.

store and forward E-mail function involving the storage of messages at an intermediate point until their addressees are ready to receive them. As an example, an E-mail message on an online service such as CompuServe is stored until the addressee signs on, at which time it is forwarded to the addressee.

sysop Someone associated with a bulletin board system (BBS) or one of the forums of an online service, and to whom others may address questions. (Originally a contraction of "system operator.") Sysops serve as guides, mentors, and suppliers of information and solutions to problems. Sysops are often voluntary and unpaid. However, manufacturer- or service-sponsored forums may have one or more paid sysops.

TD See *SD.*

text transfer See *file transfer.*

TX See *SD.*

training Process by which modems get in step with one another. If your modem has a speaker, you will hear a rushing sound during the training process. Training occurs during the initial modem *handshake*, and (with some modem types) can occur again whenever one of the modems senses a loss of synchronization.

UART Universal asynchronous receiver/transmitter. The integrated circuit used to control a serial interface on a computer (COM port), terminal, or similar device. Older UARTs can be identified by the number 8250 or 16450. The latest UARTs, bearing the number 16550 or 16550A, offer improved performance when used with the faster modems (V.32 and V.32 *bis*) with data-compression capabilities (MNP Class 5 or V.42 *bis*). A personal computer's UART (or UARTs) can be found on the motherboard, or on a serial adapter board. The Microsoft Diagnostic command (MSD), supplied with Windows (in the WINDOWS directory), will tell you which kind of UART you have on each COM port.

unbalanced circuit A circuit which shares a common return connection (signal ground) with other circuits. The number of connectors need only be one more than the number of circuits. EIA-232-D specifies only unbalanced circuits. See also *EIA-232-D.*

upload Term used customarily to describe the transfer of a file from a smaller system to a larger system (e.g, from a personal computer to an online service). May be used as a noun or a verb.

V.14 Specifies the conversion of start-stop data streams to synchronous data streams for transmission by modems conforming to CCITT Recommendations V.32 and V.32 *bis*. All V.32 and V.32 *bis* modems designed for attachment to the COM port of a personal computer incorporate this capability.

V.17 Specifies fax transmission and reception at 14,400 bps.

> The modems specified by V.21, V.22, V22 *bis*, V.32, and V.32 *bis* all operate over switched (dial-up) lines in full-duplex mode. All support start-stop (asynchronous) data streams (for which V.32 and V.32 *bis* require V.14 capability).

V.21 Specifies modems operating at up to 300 bps.

V.22 Specifies modems operating at 1200 bps and at a fallback rate of 600 bps. Not as common in the U.S. as the Bell System 212A standard.

V.22 *bis* Specifies modems operating at 2400 bps.

V.24 Specifies the interchange circuits by which a DTE and a DCE communicate with one another. See also *EIA-232-D.*

V.25 *bis* Recommends a standard for automatic calling and answering equipment for DCEs. Because of the success of Hayes-compatible modems, this standard is rarely used.

V.27 *ter* See *Group 3 fax.*

V.28 Specifies the electrical characteristics of the interchange circuits between a DTE and a DCE. This is not the latest specification, but it is the most common. See also *EIA-232-D.*

V.29 See *Group 3 fax.*

V.32 Specifies modems operating at 9600 bps, with a fallback rate of 4800 bps.

V.32 *bis* Specifies modems operating at 14,400 bps, with fallback rates of 12,000 and 7200 bps, and the two rates supported by V.32.

V.32 *terbo* Specifies modems operating at 19,200 bps, with fallback rates of 16,800 bps and the five rates supported by V.32 *bis* and V.32. This is not yet an accepted standard (or a published CCITT Recommendation) and *terbo* (a play on the word "turbo") is not a proper CCITT suffix (which would be *ter*).

V.42 Specifies an error-correction method for modems.

V.42 *bis* Specifies a data-compression standard for modems.

V.54 Specifies loopback-testing methods for DCEs and DTEs. There are four basic tests—a local digital loopback test that is used to test the DTE's send and receive circuits, a local analog loopback test that is used to test the local modem's operation, a remote analog loopback test that is used to test the communication link to the remote modem, and a remote digital loopback test that is used to test the remote modem's operation. If a modem has V.54 capability (most V.32 and V.32 *bis* modems do), its manual should include documentation on performing the various tests. Version 7 of the Norton Utilities (from Symantec) also includes a local digital loopback test for your PC's COM ports, for which you will need the optional jumper plug offered with the software. Where a modem supports local digital loopback testing, it simulates the jumper plug and does not, therefore, need to be disconnected.

V.FAST Specifies modems operating at rates up to 28,800 bps, with various fallback rates. The term V.FAST is informal and is based on the assumption that CCITT will assign a V-Series number to it.

APPENDIX D

Modem Shopping Check Lists

All terms used in this appendix may be found in the Glossary in Appendix C. Most of them are also explained in the main part of the book.

Decide Which Class of Modem You Want

Make your basic modem decision before you go shopping for it. Your three major choices are:

- 2400 bps
- 9600 bps
- 14,400 bps

Whichever of these you choose, you may want fax capability, which is not always offered with 2400-bps modems, is fairly standard with 9600-bps, and is almost inevitable with 14,400 bps modems.

COMMON SENSE

> Avoid the 2400-bps choice unless you either plan very limited use of your modem, or feel you must stay under $100.

PITFALL!

> Modem packages and modem advertisements may use baud, bps, or BPS for bits per second, and kbaud, Kbaud, kbps, Kbps, or KBPS for kilobits per second—terms that all indicate the same thing: the speed of your modem. Until the manufacturers and resellers can be persuaded to agree to a standard, you'll just have to get used to the confusion.

Another significant choice is internal versus external. Consider your preference in advance, but leave the choice until you've done your comparison shopping. (Exception: If you have no available slots in your PC, you must get an external modem.) There may be a special deal available on one kind, but not the other.

Select the Appropriate Checklist (or Lists)

There are tear-out checklists for all three modem speeds in this appendix. Unlike the rest of the book, they're not copyrighted material, so feel free to make photocopies.

COMMON SENSE

> If you're reading this book in the computer store book section, and haven't bought it yet, *don't tear out the checklists*. But unless you absolutely must buy your modem today, buy the book, take it home, and read it. If it helps you make the right decision, then it's a bargain. If you do buy your modem today, buy the book anyway. It will get you started in the right direction.

If you've decided on 9600 bps, take 14,400-bps check lists along too. You may find 14,400-bps modems that are priced lower than some 9600-bps modems. Why pass up a chance to avoid paying for an upgrade later?

By the time you read this, a new class of modem may be available—one with "soft-loadable" characteristics. Today you might buy it with a V.32, MNP Class 5, or V.42 *bis* support diskette; in six months, you might buy a fax-upgrade diskette; next year you might buy a V.32-*bis*-upgrade diskette. What would make it really worth considering, though, is if the manufacturer guarantees you'll eventually be able to upgrade it to the new standard (not yet firmly established) for 19,200-bps operation.

Miscellaneous Considerations

Here are some items you may want or need to consider. We've kept them out of the tear-out checklists to keep things simple.

- *Phone socket.* All modems have a means to connect them to the line. Most also have a socket into which you can plug a phone. But don't reject a modem just because it doesn't have a phone socket. You can always buy a splitter for your wall jack (if you don't have a spare wall jack), and plug the phone in that way. Software (such as Windows Cardfile) that places voice calls for you will work either way.

- *Manual speaker volume control.* Modem speakers are very desirable, though not essential. We've said this in the book, and included them in the checklists. Speaker phones have volume controls for obvious reasons. However, modem speakers are not used for speech, and need not be on all the time. If a modem doesn't support the Hayes speaker control commands (M0 to M3), it will turn the speaker off when the connection is made (which is the default). If the modem supports the Hayes speaker volume commands (L0 to L3), you can put the appropriate command in your initialization string. Put a manual volume control in the "nice-to-have-if-included-anyway" category.

- *Synchronous operation.* V.22 *bis*, V.32, and V.32 *bis* modems communicate with one another synchronously (or, if you're a linguistic purist, isochronously). They communicate with your computer's COM port asynchronously (that is, in start-stop mode). The more expensive V.32 and V.32 *bis* modems can also provide clocking signals to the DTE, typically to a PC-based adapter supporting the SDLC (synchronous data link control) or HDLC (high-level data link control) protocol. Unless you have a reason (usually job-related) to communicate with something like an IBM System/370, System/390, or AS/400 computer—and you have the appropriate PC-based adapter—you don't need to spend extra (usually 100% or more extra!) for synchronous DTE support.

- *Voice/data switch.* Although this is a valuable feature for synchronous operation (where autodialing is much less common), it's unnecessary for a modem that is to be attached to a PC's COM port (and not relevant for internal modems).

- *TDD capability.* TDD stands for Telephone Service for the Deaf and Disabled. Local phone companies provide terminals, at no charge, to those who qualify. If you have a friend or relative who qualifies, you may want to use your modem to communicate with him or her. In most cases, you will require a modem that supports Bell 103 operation—an attribute shown in the checklists as being (ordinarily) of questionable value. TDD communication is a clear exception to this characterization. If you want to communicate with a TDD subscriber, look for Bell 103 on the modem box. Your local phone company will provide you with the necessary setup instructions for TDD communication.

Start Your Comparison Shopping

Once you have determined that a modem possesses all the essential attributes (and reject any modem that does not), see how many of the desirable attributes it has. Then consider those, the price, the warranty, and the store's policy on returns and refunds, before making your decision.

For mail order, make sure you add shipping and handling costs to the price. You may get a break on the sales tax if the mail order company is out of state, and if your state doesn't impose a tax on out-of-state shipments.

Make Your Decision

No doubt you'll want to include some subjective judgments, like the look and feel of the packaging and of the modem itself. What about the shape of the modem? You can't inadvertently put your coffee mug on the funny-looking pyramid-shaped one, but maybe the flat one will be easier to accommodate on your cluttered desk. Maybe the salesperson is using high-pressure tactics, and makes you wary. This book and these checklists won't help you with such nonrational (but not irrational) decisions.

Once you feel comfortable with your choice, take the plunge. Nothing ventured, nothing gained.

See you online.

Checklist For Buying A 2400-bps Modem

Make several photocopies of this checklist, and use one for each modem you're considering buying. (And keep the checklist for the modem you buy with the modem manual, so you'll have a list of the specifications for your modem at your fingertips.)

Modem make/model: ______________________________

Internal ❑ External ❑

Price: ____________

Price of modem cable if not included (and make sure to buy the right one) ____________

Price of tax and/or shipping ____________

Total price: ____________

Dealer: ______________________________

Phone: ____________________

Essential Attributes

You must be able to check off all the following. If you cannot, don't buy the modem.

❑ Hayes compatibility (Hayes AT Command Language)

❑ V.22 *bis* operation at 2400 bps
(includes V.22 operation at 1200 bps)

- ❑ Bell System 212A operation at 1200 bps
  (Bell System 212 means same)

- ❑ User manual, with full specifications, setup information, and troubleshooting section.

Essential Attributes for Fax Use

If you plan to use fax software, you must be able to check off *all* the following. If you cannot, don't buy the modem.

- ❑ Group 3 fax capability
  (may be expressed as EIA-578 and/or Class 1)

- ❑ 9600-bps fax send capability

- ❑ 2400-bps fax receive capability
  (if you want to receive faxes by computer)

Desirable Attributes

General

- ❑ Price under $80

- ❑ Speaker

- ❑ On/off switch (external modem only)

- ❑ 800 number for technical support

- ❑ MNP 2, 3, 4 error correction

- ❑ V.42 error correction

- ❑ MNP 5 data compression

- ❑ V.42 *bis* data compression

- ❑ Modem cable included (external modem only)
- ❑ Communications software included

For Fax Use

- ❑ Fax software included
- ❑ 9600-bps fax send/receive capability
- ❑ Class 2 operation

Attributes of Questionable Value

- ❑ Bell System 103 operation up to 300 bps
  You'll only need this in the unlikely event that you intend to communicate with older hardcopy terminals.
- ❑ V.21 operation up to 300 bps
  You'll only need this in the *extremely* unlikely event that you intend to communicate with an older hardcopy terminal situated in Europe.

Checklist for Buying a 9600-bps Modem

Make several photocopies of this checklist, and use one for each modem you're considering buying. (And keep the checklist for the modem you buy with the modem manual, so you'll have a list of the specifications for your modem at your fingertips.)

Modem make/model: ______________________________

Internal ❑ External ❑

Price: ____________

Price of modem cable if not included (and make sure to buy the right one) ____________

Price of tax and/or shipping ____________

Total price: ____________

Dealer: ______________________________________

Phone: ____________________

Essential Attributes

You must be able to check off all the following. If you cannot, don't buy the modem.

❑ Hayes compatibility (Hayes AT Command Language)

❑ V.32 operation at 9600 and 4800 bps

- ❑ DTE (COM port) speeds up to 38,400 bps
  (needed to take full advantage of data compression)
- ❑ V.22 *bis* operation at 2400 bps
  (includes V.22 operation at 1200 bps)
- ❑ Bell System 212A operation at 1200 bps
  (Bell System 212 means same)
- ❑ MNP 2, 3, 4 error correction
  (included if MNP 5 is specified)
- ❑ V.42 error correction
  (included if V.42 *bis* is specified)
- ❑ MNP 5 data compression
- ❑ V.42 *bis* data compression
- ❑ User manual, with full specifications, setup information, and troubleshooting section.

Essential Attributes for Fax Use

If you plan to use fax software, you must be able to check off *both* of the following. If you cannot, don't buy the modem.

- ❑ Group 3 fax capability
  (may be expressed as EIA-578 and/or Class 1)
- ❑ 9600-bps fax send/receive capability

Desirable Attributes

General

- ❑ Price under $200.

- ❑ Speaker
- ❑ On/off switch (external modem only)
- ❑ 800 number for technical support
- ❑ Modem cable included (external modem only)
- ❑ Communications software included
- ❑ MNP capabilities above Class 5
- ❑ 16550 or 16550A UART (internal modem only)

For Fax Use

- ❑ Fax software included
- ❑ Class 2 operation

Attributes of Questionable Value

- ❑ Bell System 103 operation up to 300 bps
  You'll only need this in the unlikely event that you intend to communicate with older hardcopy terminals.

- ❑ V.21 operation up to 300 bps
  You'll only need this in the *extremely* unlikely event that you intend to communicate with an older hardcopy terminal situated in Europe.

Checklist for Buying a 14,400-bps Modem

Make several photocopies of this checklist, and use one for each modem you're considering buying. (And keep the checklist for the modem you buy with the modem manual, so you'll have a list of the specifications for your modem at your fingertips.)

Modem make/model: ______________________________

Internal ❏ External ❏

Price: ____________

Price of modem cable if not included (and make sure to buy the right one) ____________

Price of tax and/or shipping ____________

Total price: ____________

Dealer: ______________________________

Phone: ____________________

Essential Attributes

You must be able to check off *all* the following. If you cannot, don't buy the modem.

- ❏ Hayes compatibility (Hayes AT Command Language)
- ❏ V.32 *bis* operation at 14,400, 12,000, and 7200 bps
- ❏ DTE (COM port) speeds up to 57,600 bps (needed to take full advantage of data compression)

- ❑ V.32 operation at 9600 and 4800 bps
- ❑ V.22 *bis* operation at 2400 bps
  (includes V.22 operation at 1200 bps)
- ❑ Bell System 212A operation at 1200 bps
  (Bell System 212 means same)
- ❑ MNP 2, 3, 4 error correction
  (included if MNP 5 is specified)
- ❑ V.42 error correction
  (included if V.42 *bis* is specified)
- ❑ MNP 5 data compression
- ❑ V.42 *bis* data compression
- ❑ User manual, with full specifications, setup information, and troubleshooting section.

Essential Attributes for Fax Use

If you plan to use fax software, you must be able to check off *both* of the following. If you cannot, don't buy the modem.

- ❑ Group 3 fax capability
  (may be expressed as EIA-578 and/or Class 1)
- ❑ 9600-bps fax send/receive capability

Desirable Attributes

General

- ❑ Price under $240.
- ❑ Speaker
- ❑ On/off switch (external modem only)

- ❏ 800 number for technical support
- ❏ MNP capabilities above Class 5
- ❏ Modem cable included (external modem only)
- ❏ Communications software included
- ❏ 16550 or 16550A UART (internal modem only)

For Fax Use

- ❏ Fax software included
- ❏ 14,400 bps fax send/receive capability (V.17)
- ❏ Class 2 operation

Attributes of Questionable Value

- ❏ Bell System 103 operation up to 300 bps
  You'll only need this in the unlikely event that you intend to communicate with older hardcopy terminals.

- ❏ V.21 operation up to 300 bps
  You'll only need this in the *extremely* unlikely event that you intend to communicate with an older hardcopy terminal situated in Europe.

INDEX

C

D

E

F

G

H

I

J-K

L

M

Q-R

S

T

U

V

W-Z